PRACTICAL PROGRAM
EVALUATION

To the memory of my mother, Huang-ai Chen

PRACTICAL PROGRAM
EVALUATION

ASSESSING AND IMPROVING PLANNING, IMPLEMENTATION, AND EFFECTIVENESS

HUEY-TSYH CHEN

University of Alabama at Birmingham

SAGE Publications
Thousand Oaks ▪ London ▪ New Delhi

For information:

Sage Publications, Inc.
2455 Teller Road
Thousand Oaks, California 91320
E-mail: order@sagepub.com

Sage Publications Ltd.
1 Oliver's Yard
55 City Road
London EC1Y 1SP
United Kingdom

Sage Publications India Pvt. Ltd.
B-42, Panchsheel Enclave
Post Box 4109
New Delhi 110 017 India

Printed in the United States of America

Library of Congress Cataloging-in-Publication Data

Chen, Huey-Tsyh.
Practical program evaluation: Assessing and improving planning, implementation, and effectiveness / Huey T. Chen.
    p. cm.
Includes bibliographical references and index.
ISBN 0-7619-0232-5 (cloth)—ISBN 0-7619-0233-3 (pbk.)
  1. Evaluation research (Social action programs) I. Title.
H62.C3647 2005
300.′ 72—dc22

                        2004011272

04   05   06   07   10   9 8 7 6 5 4 3 2 1

Acquisitions Editor:	Lisa Cuevas Shaw
Editorial Assistant:	Margo Beth Crouppen
Production Editor:	Melanie Birdsall
Copy Editor:	Liann Lech
Typesetter:	C&M Digitals (P) Ltd.
Proofreader:	Mary Meagher
Indexer:	Kathy Paparchontis
Cover Designer:	Edgar Abarca

CONTENTS

PREFACE

———•◦•———

When I was a professor at the University of Akron, I had a commitment to write a practical evaluation book, based upon the theory-driven approach, for students in an advanced evaluation class and for evaluation practitioners. The majority of the manuscript was finished there before I accepted a position for heading an evaluation branch at the Centers for Disease Control and Prevention (CDC) in the fall of 1997. The new job had mixed impacts on the book. On one hand, it created a long delay in completing the book. On the other hand, the book ultimately benefitted from the experience of practicing evaluation on a day-to-day basis and working closely with many evaluation practitioners inside and outside the agency. The CDC experience motivated me to finish the manuscript soon after I returned to the academic world in the fall of 2002 at the University of Alabama at Birmingham. I had been on a long journey and had made a tremendous effort to complete the book, but the intellectual growth accompanying the progress of the book has been invaluable and memorable to me.

In working on this book, I have found that many evaluation concepts may be too vague or ambiguous for practical use. In my experience, the vagueness and ambiguity of these concepts create difficulties for evaluation practitioners in conceptualizing problems, selecting and applying evaluation methods, and communicating with stakeholders. For example, the concepts such as formative research, formative evaluation, process monitoring, and process evaluation have been lumped together as process evaluation in the existing literature. However, in practice, these evaluation approaches are quite different in terms of premises on which they are based, what kind of data are collected, how to collect the data, and how to utilize the information. Without distinction, stakeholders might want one thing, but evaluation practitioners provide another. In order to clarify evaluation concepts for practical use, this book creates many phrases that are not commonly seen in the current literature. It is my hope that readers will not simply regard these concepts and phrases as jargons or burdens but rather as useful tools for sharpening their understanding of evaluation, enriching their evaluation toolbox, and applying evaluation correctly and fruitfully.

I am greatly indebted to Debra Rug and Len Bickman for their encouragement, suggestions, and help on the book. I appreciate Mel Mark for providing informative suggestions for revising the book. I am grateful to Stewart Donaldson, Katrina Bledsoe, John Gargari, and Thomas Creger for reviewing and commenting on the manuscript. I am also grateful to Lisa Cuevas Shaw and Melanie Birdsall from Sage for providing all the necessary assistance for publishing this book. Finally, support and encouragement from my wife, Jean, my daughter, Charlene, and my son,Victor, were essential for completing this book. Readers may fax their comments and questions on the book to (205) 934-9325, or mail them to the Department of Health Behavior, The University of Alabama at Birmingham, 1530 3rd Avenue South, RPHB-227K, Birmingham, AL 35294-0022.

OVERVIEW OF THE BOOK

———•◦•———

This book is about program evaluation in action. More specifically, the book applies the theory-driven evaluation approach to address the following three steps that are vital to the practice of program evaluation.

1. *Systematically Identifying Stakeholders' Needs.* Almost always, program evaluation is initiated in order to meet the particular evaluation needs of a program's stakeholders. If a program evaluation is to be useful to those stakeholders, it is *their* expectations that evaluators must keep in mind when designing the evaluation. Stakeholders' evaluation needs vary at different program stages. For example, evaluation needs during the planning and development of a program are quite different from evaluation needs for a mature program. The precise communication and comprehension of stakeholder expectations is crucial; to facilitate the communication process, this book presents a comprehensive framework for the effective identification of evaluation needs. The framework covers the gamut of evaluation needs, from program planning through implementation to outcomes.

2. *Selecting Evaluation Options Best Suited to Particular Needs (and Reconciling Trade-offs Among These Options).* Once the stakeholders' expectations are identified, evaluators must select a strategy for addressing each evaluation need. Many options are available. The book discusses them, exploring the pros and cons of each and acknowledging that trade-offs sometimes must be made. Furthermore, it suggests practical principles that can guide evaluators to the best choice in evaluation situations they are likely to encounter.

3. *Putting the Selected Approach Into Action.* Using illustrative examples from the field, the book details the methods and procedures involved in the various evaluation options. How does the program evaluator actually carry out an evaluation in an effort to meet real evaluation needs? Here, practical approaches are discussed; and yet this book avoids becoming a "cookbook." The principles and strategies of evaluation that it presents are backed by theoretical justifications, which are also explained. This context, it is hoped, creates the latitude, knowledge, and flexibility with which program evaluators can design suitable evaluation models.

The book contains four parts. Part I discusses the basic concepts and conceptual framework used in the book. Part II discusses how program evaluation can be used to assist stakeholders as they plan programs. Part III focuses on evaluation approaches and methods that are useful in fine-tuning and stabilizing the implementation process of fledgling programs. Finally, Part IV describes how to assess program implementation once programs mature.

PART I

INTRODUCTION

The three chapters in Part 1 of this book provide general information about the foundations and the application of program evaluation principles. Basic ideas are introduced, and a conceptual framework is presented. The first chapter sets out the purposes of the book and discusses the nature, characteristics, and strategies of program evaluation. The second chapter introduces concepts of the *program theory* that grounds many of the guidelines found in the book. Finally, in Chapter 3, program evaluators will find a systematic taxonomy comprising various evaluation approaches to choose among when faced with particular evaluation needs.

ONE

FUNDAMENTALS
FOR PRACTICING
PROGRAM EVALUATION

———•◦•———

THE NATURE AND CHARACTERISTICS OF PROGRAMS

This book is all about program evaluation—the evaluation of "programs." But what is meant by that term? How are the nature and characteristics of programs related to evaluation? The programs that evaluators can expect to evaluate have different names, such as *treatment* program, *action* program, or *intervention* program, and come from different substantive areas, such as health promotion, education, criminal justice, and welfare. Nevertheless, they all share the common feature of being organized efforts to enhance human well-being—whether by preventing disease, reducing poverty, or teaching skills. For convenience, programs and policies of any type are usually referred to in this book as "intervention programs," or simply "programs." Program evaluation is defined as the application of evaluation approaches, techniques, and knowledge to systematically assess and improve the planning, implementation, and effectiveness of programs.

The terminology of systems theory provides a means of illustrating, rather easily, the nature and characteristics of a given program. A program must perform two functions in order to succeed and survive. First, internally, it must ensure the smooth transformation of inputs into desirable outputs. For example, any education program would be in big trouble if faced with disruptions like

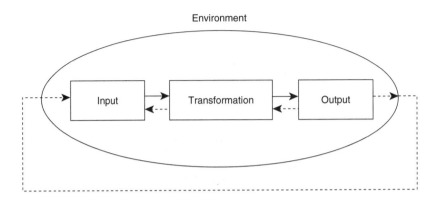

Figure 1.1   A System View of a Program

high staff turnover, excessive student absenteeism, or insufficient textbooks. Second, and externally, a program needs to continuously interact with its environment in order to obtain the resources and support necessary for its survival. That same education program would also become quite vulnerable if support from parents and school supervisors evaporated. Thus, because programs are subject to the influence of their environments, every program comprises an "open system." Drawing further upon systems theory, any intervention program can be conceptualized as having the following five components (input, transformation, output, feedback, and environment),[1] illustrated in Figure 1.1.

Inputs. Among the components of an intervention program are "inputs" from the environment. Inputs are resources taken in from the environment. These can include finances, technology, equipment, facilities, personnel, and clients. Inputs form and sustain a program, but they cannot work effectively without systematic organization. Usually, a program requires an implementing organization that can secure and manage its inputs.

Transformation. The component called "transformation" represents the processes by which a program converts inputs into outputs. Transformation begins with the initial implementation of the treatment/intervention prescribed by a program. Transformation is, then, the stage during which implementers provide services to clients. For example, the implementation of an new curriculum in a school may mean the process of teachers teaching students new subject matter in accordance with existing instructional rules and administrative guidelines; this would represent transformation. Transformation also includes those sequential events necessary to achieve desirable outputs. For example, in order to increase students' math and reading scores, an education program may need to first boost students' motivation to learn.

Outputs. "Outputs" are the results of transformation. One crucial output is attainment of the program's goals, which alone justifies the existence of the program. For example, an output of a treatment program for those engaging in spousal abuse is whether or not the abuse ends.

Environment. The "environment" refers to any factors that, despite lying outside a program's boundaries, can nevertheless foster or constrain that program's implementation. Such environmental factors may include social norms, political structures, the economy, funding agencies, interest groups, and concerned citizens. Because an intervention program is an open system, it depends on the environment for its inputs: clients, personnel, money, and so on. Furthermore, the continuation of a program often depends on the way the general environment perceives program outputs. Are the outputs valuable? Are they acceptable? A day care program provides an example. If its staff is suspected of abusing children, the environment would find that output unacceptable. Parents would immediately remove their children from the program; the community might press criminal charges or at least boycott the program.

Feedback. To succeed—to correct any problems or adjust course effectively—an open system requires information about inputs and outputs, transformation, and the environment's responses to these components. This information is called "feedback." Feedback is what program evaluation is all about. Programs need information to gauge whether inputs are adequate and organized, interventions are implemented appropriately, target groups are reached, and clients receive quality services, and outputs demonstrate the attainment of goals and meeting of funding agencies' and decision makers' (e.g., stakeholders') expectations. Without feedback, a system flies blind and is bound to deteriorate and eventually die. Without insightful program evaluation, programs fail. The action of feedback within the system is indicated by the dotted lines in Figure 1.1.

The five components of an open system can also be identified in any given "policy," which is a concept closely related to a program. Although policies may seem grander than programs—in terms of the envisioned magnitude of intervention, the number of people affected, and the legislation process—the principles and issues this book addresses are relevant to both. Throughout the rest of the book, in fact, the word *program* may be understood to mean program *or* policy.

THE NATURE AND CHARACTERISTICS OF EFFECTIVE EVALUATION

The transformation of inputs into outputs, and the interaction between program and environment, are nothing if not dynamic. Their fluctuations often make the evaluation of a program as challenging as it is necessary. The program evaluator can help ensure the quality of feedback about a program by making certain that a program evaluation is *future action-directed,* carries both scientific and stakeholder *credibility,* and takes a *holistic approach.*

Future Action-Directedness

One popular view of program evaluation is the assessment of the merits of a program. However, merit assessment is just one possible approach to program evaluation; often, program evaluation needs to go beyond merit assessment. When actually practicing an evaluation, all program evaluators quickly learn that stakeholders are eager to figure out what to do next. Stakeholders find evaluations useful if they both offer conclusions on how well programs have worked *and* provide information that assists the stakeholders in figuring out what must be done next to securely attain—or even surpass—program goals. Thus, the assessment of a program's performance or merit is only one part of program evaluation (or, alone, provides a very limited type of evaluation). To be most useful, program evaluation needs to equip stakeholders with knowledge of those program elements that are working well and those that are not. Program evaluation in general should facilitate stakeholders' search for appropriate actions to take in addressing problems and improving programs. There are important reasons why evaluations must move beyond narrow merit assessment into the actual determination of needed improvements. Just as, in the industrial and business world, information on product improvement is provided by research-and-development engineering and by market research, in the world of intervention programs, the agency or organization overseeing an effort relies on program evaluation to help it continually guarantee or improve the quality of services provided.

Consider that those intervention programs typically operate in the public sector. In the private sector, the existence or continuation of a product is usually determined by market mechanisms. That is, through competition for consumers, a good product survives and a bad product is forced from the market: Consumers reject the dissatisfactory product and choose a better quality alternative. However, the great majority of action programs do not encounter any market competition (Chen, 1990). Drug abusers in a community may find, for example, that there is only one treatment program available to them. In the absence of an alternative, the treatment program is likely to continue whether or not its outcomes justify to do so. Furthermore, well-known good-intention programs, such as Head Start, would not be discontinued based on an evaluation saying the programs were ineffectual; decision makers rarely use program evaluation results alone to decide whether a program will go on.

Under these circumstances, an evaluation that simply assesses the merit of a program's past performance and cannot provide stakeholders with insights to help them take the next step is of limited value (Cronbach, 1982). In fact, many stakeholders look to a broad form of program evaluation to point out apparent problems, as well as strengths upon which to build. In general, in order to be responsive and useful to stakeholders, program evaluation should meet both assessment needs *and* improvement needs rather than confine itself solely to merit assessment. Stakeholders need to know if the program is reaching the target group, if treatment/intervention is being implemented as directed, if staff are providing adequate services, if clients are making a commitment to the program, and if the environment seems to be helping or hindering the delivery of services. Whereas any part of this information can be difficult for stakeholders to collect, program evaluators have the necessary training and skills to gather and synthesize it all systematically.

Merit assessment remains an important approach or activity within program evaluation. In a broad sense, however, assessment is a means, rather than the end, of program evaluation. Our vision of program evaluation should extend beyond the design of supremely rigorous and sophisticated assessments. It is important to grasp that evaluation's ultimate task is *to produce useful information that can enhance the knowledge and technology we employ to solve social problems and improve the quality of our lives.*

Scientific and Stakeholder Credibility

The scientific credibility of a program evaluation reflects the extent to which that evaluation was governed by scientific principles. Typically, in scientific research, scientific credibility is all that matters: The more closely research is guided by scientific principles, the higher its credibility. However, scientific credibility is just the first kind of credibility that program evaluation must establish. As an applied science, program evaluation also exhibits varying degrees of *stakeholder* credibility. Stakeholders are those who have vested interests in the evaluation. The stakeholder credibility of a program evaluation reflects the extent to which stakeholders believe the evaluation's design gives serious consideration to their views, concerns, and needs. In general, the more an evaluation responds to stakeholders' views, concerns, and needs, the higher its stakeholder credibility will be.

The ideal evaluation achieves both high scientific and high stakeholder credibility. And yet the two do not go automatically hand in hand. An evaluation can have high scientific credibility but little stakeholder credibility, as when evaluators follow all the scientific principles but set the focus and criteria of evaluation simply as they see fit. Their evaluation is easily dismissed or rejected by stakeholders, despite scientific credibility, because it fails to reflect the stakeholders' intentions for the program. For example, there are good reasons for African-Americans to be skeptical of scientific experiments without community inputs, due to incidents such as the Tuskegee syphilis experiment (Jones, 1981). Researchers in the experiment withheld effective treatment from African-American men suffering from syphilis so that long-term effects of the disease could be documented. Conversely, an evaluation overwhelmed by the influence of stakeholders such as program managers and implementers may neglect its scientific credibility, ending in evaluation results that are suspect.

The challenge in program evaluation is to achieve a balance of both kinds of credibility. A helpful strategy is to pursue stakeholder credibility in the earliest phases of evaluation design but to yield to scientific principles later in the process (Chen, 1990). Initially, evaluators experience a great deal of interaction and communication with a program's stakeholders, for the specific purpose of understanding their views, concerns, and needs. Evaluators then incorporate the understanding they have acquired into the research focus, questions, and design, along with the necessary scientific principles. From this point on, to establish scientific credibility, the evaluators require clear autonomy to design and conduct evaluations without interference from stakeholders. Stakeholders are usually receptive to this strategy, especially when evaluators explain the procedure to them at the very beginning of the process (Perry & Backus, 1995). While stakeholders do not object to a program being evaluated, or dispute the evaluator's need to follow scientific procedures, they do expect the evaluation to be fair, relevant, and useful (Chen, 2001).

Holistic Approach

Many disciplines or areas of inquiry have an evaluative ingredient. Some disciplines gauge consumer satisfaction with products (product evaluation). Some assess an employed individual's performance in a company or organization (personnel evaluation). Some weigh a person's skills or qualifications (college entrance examinations). Some assess the effects of new drugs (biomedical experimentation). Some judge performance in a sport (competitive figure skating or diving). Some assess the appropriateness of the handling of money (public accounting). The nature of a program, though, as discussed earlier in this chapter, suggests that programs are clearly distinct from the objects of these other disciplines, and thus both the nature of a program evaluation *and* its associated principles and strategies belong to it alone.

One important characteristic distinguishing program evaluation is its need, rarely shared by other disciplines, to use a holistic approach to assessment. The holistic approach to assessment finds it imperative to include contextual or transformation information when assessing the merit of a program. By comparison, product evaluation is more streamlined: It may get away with an exclusive focus on the intrinsic value of its object. Products like televisions can be assessed in terms of intrinsics such as picture, sound, durability, price, and so on. In many situations, however, the value of a program may be ecological as well as intrinsic or inherent. That is, to adequately assess the merit of a program, both its intrinsic value and the context in which that value is assigned must be considered together. Take as an example the case of an educational program that, according to strictly performance-based evaluation, has attained its goals (which are its intrinsic values). But in what context was the performance achieved? Perhaps goals were attained by "creaming," which is the deliberate admission of only those students expected to meet with success in the program and the accompanying rejection of students not so clearly destined to succeed but perhaps more in need of the program. Does the program's performance still deserve loud applause? Probably not.

Similarly, what about a case in which program success is due to participation of a group of highly talented, well-paid teachers with ample resources and strong administrative support, but the evaluated program is intended for use in ordinary public schools? This "successful" program may not even be relevant, from the viewpoint of the public schools, and is not likely to solve any of their problems. In program evaluation, *how* a program achieved its goals is as important as *whether* it achieved them. For example, an outcome evaluation of one family planning program in a developing country limited its focus to the relationship between program inputs and outputs; it appeared possible, on this basis, to claim success for the program. A large drop in the fertility rate was indeed observed following the intervention. Transformation information, however, showed it was misleading to make such a claim. Although the drop in fertility was real, it had little to do with the intervention. A larger factor was that, following implementation, a local governor of the country, seeking to impress his prime minister with the success of the program, actually ordered soldiers to seize men on the streets and take them to be sterilized. An evaluator with an approach that was less than holistic might have declared that the goals of the program were attained, whereas all around were people whose personal knowledge led them to condemn the program as inhumane. Lacking a holistic orientation, program evaluation may reach very misleading conclusions.

THE UNIVERSALIST VIEW OF THE
PRACTICE OF EVALUATION (AND ITS LIMITATIONS)

The approaches to and methods of evaluation that have been developed within program evaluation are plentiful (Rossi, Lipsey, & Freeman, 2004; Shadish et al., 1991). An important issue for evaluation practitioners is how to select an approach or method for their evaluation. There are two general points of views about making this determination: the universalist and the contingency. In the following section, the universalist view is discussed and its limitations reviewed, providing the basis for the next section's discussion of the contingency view.

The *universalist view* insists on the universal superiority of certain evaluation approaches and methods over others; in light of this precept, evaluators must strive to apply the endorsed "best" evaluation approaches and methods in undertaking any evaluation. The research methods involved in program evaluation provide an example. The universalist view argues that there is always a "best" research method. For instance, some evaluators advocate quantitative methods (such as randomized experiments) for use in program evaluation, whereas others insist that qualitative methods, or mixed methods, are best. Whatever his or her choice, the evaluator who advocates one best method takes the universalist view of program evaluation research methods. Formative versus summative evaluation provides a similar example. Evaluators who argue that summative evaluation is better than formative evaluation—as well as those who argue the opposite— demonstrate universalist principles for performing evaluation. One reason for the popularity of the universalist view (notably during program evaluation's infancy) is that universalist principles appear straightforward, forceful, and easy for practitioners to follow. Another reason is the convenience the universalist view provides to evaluation theorists seeking to frame their arguments, highlight advantages of their approaches, and attract attention.

In reality, though, particular programs vary considerably with regard to structure (e.g., fluid vs. routine, large scale vs. small scale) as well as to expressed evaluation needs, availability of pertinent data, amount of funding earmarked for evaluation, and other realities. Given the plethora of such variables, despite the universalist view, it seems questionable at the very least whether there is one, best evaluation approach or research method covering every evaluation. The application of the universalist view and ensuing principles makes evaluators vulnerable to the failures of an indiscriminate, one-size- fits-all type of evaluation.

THE CONTINGENCY VIEW AND
THE PROGRAM THEORY PERSPECTIVE

An alternative point of view from which to conceptualize, develop, and apply program evaluation principles, approaches, and methods is the *contingency view,* and it postulates that there is no single best way to conduct program evaluation; rather, the choice of approaches and methods for program evaluation should be situational. The basic principle of the contingency view is that the individual natures of programs and the uniqueness of evaluation purposes and contextual circumstances require use of a range of evaluation approaches and methods. For example, taking a contingency view, Chen (1996) specified the conditions (e.g., depth of information, credibility of data, breadth of program boundaries) under which quantitative, qualitative, and mixed methods will prove most serviceable to program evaluators. From the contingency viewpoint, guidelines for evaluation may be less straightforward than under the universalist view, but this creates the advantage of avoiding simplistic assumptions. In general, the contingency view appears to fit better with the reality in which practitioners currently conduct even state-of-the-art program evaluations.

One key evaluation perspective incorporating the contingency viewpoint in its principles and strategies is the *program theory evaluation perspective,* often called simply the program theory perspective (Chen, 1990, 1994, 1996, 1997). The program theory perspective has benefited greatly from the intensively discussed topics comprised by theory-driven or theory-based evaluation (e.g., Bickman, 1987a, 1987b, 1990; Chen & Rossi, 1980, 1992; Connell et al., 1995; Donaldson, 2003; Fulbright-Anderson et al., 1998; Pawson & Tilly, 1997; Rogers et al., 2000; Suchman, 1967, 1969; Weiss, 1997, 1998; Wholey, 1987). The program theory perspective can help guide practitioners or students toward an understanding of the circumstances in which a particular set of evaluation approaches and methods is appropriate for evaluating a particular program.

The program theory perspective calls the practitioner's attention to the following major factors influencing selection of evaluation approaches and methods: (a) Which stage or stages of the program cycle will be the focus of the evaluation? (b) What do stakeholders want from the evaluation— assessment-oriented information, improvement-oriented information, or both? (c) What evaluation options potentially fit the given program's environmental and other circumstances, as well as stakeholders' needs? (d) What trade-offs among these options will be most profitable? The program theory perspective provides a conceptual framework that practitioners can use to find effective answers to these four questions. Detailing this conceptual framework is one of two major purposes of this book. The second purpose is to provide systematic advice on how to fruitfully conduct a program evaluation once an evaluation approach is identified. The program theory perspective and its conceptual framework are discussed at length in Chapter 2.

AUDIENCES AND USES OF THIS BOOK

This book was prepared for two audiences. The first is evaluation practitioners, especially those who seek new knowledge to strengthen their practical skills or expand the scope of their work. Such practitioners should generally look to the book to broaden their vision of evaluation alternatives, enhancing their skills for designing evaluations fitting a variety of program circumstances and evaluation purposes. Seasoned program evaluators may find in this book both valuable insights into established evaluation strategies and approaches, and new ideas for practice. The second anticipated audience is students, possibly students who completed "Evaluation 101," but certainly those interested in issues of the actual practice of evaluation, including difficulties evaluators can expect and practical means of dealing with them. The book may liberate such students from the notion that evaluations are mainly methodological activities. This should help prevent evaluation students from feeling like mindless number crunchers. The book could even challenge them to seek strategies for broadening basic social science theories learned in the classroom, linking these to action and intervention theories employed in the field by program staff, evaluators, and social reformers.[2]

Chapter 2 discusses the concepts and conceptual framework of the program theory perspective; these provide the foundation for the materials in the rest of the book. In Chapter 3, a road map of evaluation options—the "evaluation taxonomy"—is presented. The evaluation taxonomy can guide evaluators and stakeholders in selecting the approaches and methods best suited to a program's circumstances and the stakeholders' needs at different program stages (program planning, initial implementation, mature implementation, and outcome), as discussed in Chapters 4 through 10.

This book can be applied to start-up programs or established programs. For a start-up program, evaluators may be asked to evaluate one or more program stages, choosing among the planning, initial implementation, mature implementation, and outcome stages. For an established program, evaluators typically are invited to conduct evaluation activities at the mature implementation stage and/or the outcome stage.

The theoretical and methodological roots of this book could be traced back to my book, *Theory-Driven Evaluations* (Chen, 1990). Readers who are interested in program theory and theory-driven evaluation are encouraged to read this book.

NOTES

1. It is important to note that, although some components identified in systems theory bear the same labels as components found in the logic model, the definitions of each are quite distinct.

2. Chapter 3 includes a more intensive discussion of the concepts of these program stages.

TWO

A CONCEPTUAL FRAMEWORK OF PROGRAM THEORY FOR PRACTITIONERS

The benefits that program theory provides to evaluation are well documented in the evaluation literature. For example, Bickman (1987a) discussed the usefulness of program theory for improving the generalizability of evaluation results, contributing to social science theory, uncovering unintended effects, and achieving consensus in evaluation planning. Weiss (1998) noted as an advantage of program theory the provision of early indications of program effectiveness. She also found program theory helpful for explaining the occurrence of program effects, which enhances the relevance of evaluation. Nevertheless, Weiss (1997) indicated that program theory's lack of clarity of the concept creates an obstacle for advanced theory-based evaluation. In Chapter 2, the book seeks to explain, for evaluation practitioners, the concepts and conceptual framework of program theory in user-friendly terms. Understanding these concepts should pave the way to using them effectively in evaluation.

For users of the book, a vital function is served by the introduction, in this chapter, of the conceptual framework of program theory. The framework helps readers grasp how the evaluation taxonomy can fruitfully guide the choice of evaluation approach or method (Chapter 3). Knowledge of the framework also elucidates the how-to of applying the various approaches and methods (Chapters 4-10).

THE PROGRAM THEORY CONCEPTUAL FRAMEWORK

One popular definition of program theory arises from causal relations. Bickman (1987a), for example, defined program theory as "a plausible and sensible model of how a program is supposed to work" (p. 5). Similarly, Weiss (1995, p. 66) viewed program theory as the picture of how and why programs work (or don't). Another popular way to understand program theory is from the context of the logic model. For example, Wholey (1987) said program theory incorporates "program resources, program activities, and intended program outcomes, and specifies a chain of causal assumptions linking program resources, activities, intermediate outcomes and ultimate goals" (p. 78). A broader definition of program theory subsuming the existing definitions was given by Chen (1990), who described program theory as "a specification of what must be done to achieve the desirable goals, what other important impacts may also be anticipated, and how these goals and impacts would be generated" (p. 43). This chapter of the book elaborates on this latter definition of program theory.

The design and implementation of an intervention program are usually based on a set of explicit or implicit assumptions by stakeholders about what action is required to solve a social problem and why the problem will respond to this action. The analysis of the explicit and implicit assumptions underlying a program is called *program theory.* Chen's definition of program theory suggests its simultaneously prescriptive and descriptive nature, a status requiring program theory to be action-oriented. Thus, program theory goes beyond typical scientific theories—those from the social and behavioral sciences, for instance—that focus solely on providing causal explanations of phenomena. Program theory can be viewed, then, as a configuration of the *prescriptive and descriptive* assumptions held by stakeholders and thus underlying the programs stakeholders create.

Descriptive Assumptions

Within program theory, descriptive assumptions concern the causal processes begetting whatever social problem a program tries to address. As an illustration, consider a treatment program for spouse abusers. According to program designers' *descriptive assumptions,* spouse abuse typically results, at least in part, from the abuser's lack of skill in dealing with anger or frustration and lack of knowledge of the law's stance on domestic violence. In light of these descriptive assumptions, the treatment program might be designed to employ counseling to develop anger management skills. It might also stress the legal consequences of committing domestic violence. The causal process underlying this treatment program's effectiveness, then, would be the instillation of fear of consequences, to encourage practice of the skills taught, to then reduce the abuse.

Assumptions about causal processes through which treatment or intervention is supposed to work are crucial for any program, because *its effectiveness depends on their validity.* If invalid assumptions dictate the strategies of a program, it is unlikely to succeed. For example, among those enrolled in the hypothetical spouse abuse treatment program, if the major motive of the abuse is belief in the patriarchal structure of families, rather than uncontrolled anger or ignorance of consequences, then the program's emphasis on anger management is unwarranted. The set of descriptive assumptions made about causal processes underlying intervention and its outcomes constitutes the causative theory (Chen, 1990) of programs. Outside the field of program evaluation, however, this phrase may not communicate well—and remember that stakeholders come from other fields. The set of descriptive assumptions can also be termed the *"change model,"* for purposes of effective communication, and throughout this book change model is substituted for *causative theory* or *descriptive theory.* The change model is emphasized in much of the theory-driven or theory-based evaluation literature (e.g., Donaldson, 2003; Weiss, 1998). As will be discussed in Chapters 3 and 4, the change model concept is very useful for developing a program rationale.

Prescriptive Assumptions

Turning now from descriptive assumptions to prescriptive assumptions, the latter are equally significant, according to program theory, in an intervention program. The prescriptive assumptions of program theory prescribe those components and activities that the program designers and other key stakeholders see as necessary to a program's success. Program designers' prescriptive assumptions thus direct the design of any intervention program. They determine the means of implementing and supporting the intervention so that the processes described in the change model can occur. Because prescriptive assumptions dictate what implemented components and activities will be required to activate the change model, they are collectively referred to as the normative theory, or prescriptive theory, of programs (Chen, 1990). But again, stakeholders (and evaluation practitioners, too) may appreciate the directness of an alternative term, *action model,* which is used in the remainder of this book. As will be discussed in Chapters 3 and 4, the action model concept is very useful for developing a program plan.

Program evaluators look to the action model for the requisites of a program, as well as for the feasibility of these requisites in the field. In the action model are found the bases for answering questions such as the following: What are the crucial elements of the intervention? What kind of organization is needed to deliver the services? Who is best qualified to deliver them? How will implementers be trained? What is the target group? How will the target group be reached?

Again, as an example, take the spouse abuse treatment program. Suppose its designers decide that the target group should be abusers convicted by a court; this decision is based on an assumption that most spouse abusers end up in court, and on the court's agreement to use the treatment program as part of an abuser's sentence. The arrangement would certainly guarantee the program a steady source of clients. It would also necessitate establishment of an administrative linkage between the court and the program's implementing organization, based on an assumption that clear channels of communication will keep the court apprised of any client's failure to attend treatment. Suppose the program designers choose group counseling as the treatment for the abusers, headed by a trained and experienced professional facilitator. This decision could stem from the program directors' favorable experiences with group therapy in other situations. Perhaps the designers decide that group counseling should be provided weekly for 10 weeks because they believe that 10 counseling sessions is a sufficient "dose" for most people. From these assumptions comes the need for the program to hire two professional counselors available for 10 consecutive weeks.

The action model deals with nuts-and-bolts issues, which are not a major topic in most modern social science theory, perhaps due to the social sciences' emphasis on developing generalizable propositions, statements, and laws. "How-to" program issues tend to be trivialized by contemporary social science theory. Plus, the action model has no proposition-like format resembling that defined by and familiar to modern social scientists. However, it is interesting to note that many classic social science texts discuss both descriptive and prescriptive theories. Both Max Weber (1947) and Emile Durkheim (1965) intensively discussed not just explanations of organizational and societal phenomena but also steps for improving organizations and societies.

The action model translates the abstract ideas that theoretically justify a program into the systematic plan necessary to organize its day-to-day activities. Implementation of the action model puts a program in motion. And just as with the change model, if the action model is based on invalid assumptions and is thus poorly constructed or unrealistic, the program is not likely to meet with success. Another example shows how important an accurate action model is to a program. The government of a developing country found that many farmers could not afford to buy fertilizer or modern equipment to increase productivity. It moved to set up low-interest loans for the farmers. Designers of this financial program postulated a particular change model: Lack of access to capital limits farmers' ability to improve productivity, and farmers would apply for low-interest loans, if available, to buy machinery and fertilizer to boost their land's productivity and their earnings. The designers' programmatic model stipulated use of the government's own banks to process applications and conduct subsequent transactions. The underlying assumption was that, as part of the government system, these banks would require simply an administrative order to diligently and responsibly implement the program; in addition, operational costs would be much less than if commercial banks became involved.

A couple of years after the program had been launched, few farmers had received loans and benefited from the program. Why? Because certain assumptions of the action model were wrong. Local staff of the government bank did not see the new program as all in a day's work. To them, the program meant another burden in addition to their already heavy workload, with no increase in rewards. Consequently, the staff members' behavior concerning implementation of the program was not quite what decision makers had assumed it would be. Not only were they unenthusiastic about the program, but they also pulled up older rules and regulations to actively discourage farmers from applying for, or to disqualify them from receiving, the loans. This maintained their accustomed workload—and made the new program fail.

In this chapter, Chen's (1990) initial conceptual framework of program theory is broadened and altered to increase its relevance within evaluation practice. The components of a change and action model are discussed as follows.

COMPONENTS OF THE CHANGE MODEL

The components of a change model are its goals and outcomes, its determinants, and the interventions or treatments it is to implement. These change model components and their interrelationships are introduced here.

Goals and Outcomes

Goals reflect the desire to fulfill unmet needs, such as with poor health, inadequate education, or poverty. Program goals are established in light of certain major assumptions about them, such as their likelihood of being well understood and supported by staff and other stakeholders; their power to motivate commitment of resources and effort; and/or their accurate reflection of stakeholders' aims in valid, measurable outcomes. A program's existence is justified through the meeting of its goals, which are usually articulated in very general, highly laudatory language in an effort to win broad support for the program. In contrast, outcomes are the concrete, measurable aspects of these goals. For example, one goal of welfare reform is to reduce dependency on welfare. An outcome linked to this goal might be increased numbers of welfare recipients obtaining jobs, alleviating need for government support. "Reducing dependency on welfare" is a notion with many ramifications; it is imprecise. But the outcome "obtaining jobs" gives specific meaning to the program's orientation.

Outcomes themselves may have components, and some outcomes may have both short-term and long-term manifestations. For example, in an HIV prevention program, the outcome over the short term may be increased use of condoms by a high-risk population. The outcome of the same program in the long term may be a lower number of HIV transmissions. Furthermore, a program's outcomes may include intended and unintended developments. If program stakeholders and evaluators suspect that unintended outcomes exist (whether desirable or undesirable), then the evaluation should include the identification of all unintended outcomes.

Determinants

To reach goals, programs require a focus, which will clarify the lines their design should follow. More specifically, each program must identify a leverage mechanism or cause of a problem upon which it can develop a treatment or intervention to meet a need. The assumption is that, once the program activates the identified leverage mechanism, or alleviation of the cause of a problem, its goals will soon be achieved. That leverage mechanism is variously called the *mediating variable,* the *intervening variable,* or the *determinant,* and in this book, the latter term is used. In the field of health promotion, theories suggest a variety of determinants that program designers and key stakeholders can deploy in a program (Bartholomew, Parcel, Kok, & Gottlieb, 2001). For example, the health belief model (e.g., Strecher & Rosenstock, 1997) outlines these determinants influencing an individual's course of action (or inaction) for a health problem: perceived susceptibility to the problem, perceived seriousness of the problem's consequences, perceived benefits of a specific action, and perceived barriers to taking action. Similarly, social learning theory (Bandura, 1977) cites self-efficacy—or the conviction that one can, in fact, carry out the behavior that elicits the outcome—as the most critical determinant of behavioral change. The PRECEDE-PROCEED model (Green & Kreuter, 1991) identifies predisposing factors, reinforcing factors, and enabling factors as important determinants for health behavioral change. The determinants identified by scientific theories are intensively studied and applied in scientific research.

Of course, few programs designed and conducted by stakeholders are designed for strict conformity to social science theory. Naturally, what is identified as the determinant often relates to the program designers' understanding of what actually causes the problem they want to alleviate and on which exact cause or causes they want a program to focus. There have been program designers, for example, who believed that urban school students' poor test performance stemmed from a lack of parental involvement, making parents the appropriate focus for programs meant to improve scores. These program designers saw in parental involvement the determinant to help students perform better; for them, it followed that, if the program activated parental involvement, student scores would improve. With a determinant identified, they could move on to figuring out how parents could be encouraged to participate and trained to help children study. Again, a program's identified determinant will provide its focus.

Social problems often have roots in multiple causes, but an intervention program usually focuses on one, or perhaps a few, determinants that program designers see as the major cause of the problem—or the most feasible to address or the one best suited to their expertise. It would be difficult for a program to deal simultaneously with all potential determinants, given typical constraints on resources and time. The unmanageability of multiple determinants aside, it remains important to specify clearly on what determinant a program has selected to focus and to justify that selection. Consider the case of juvenile delinquency in a community. High rates of such delinquency may be the result of peer pressure, failure in school, a lack of positive role models, a lack of discipline, a subculture of violence, or a dearth of economic opportunity. A program to lower rates of juvenile delinquency must state plainly, to stakeholders and the community, the cause or causes it assumes to be most relevant and the determinant or determinants upon which it will focus. Selection of the determinant or determinants could be shaped by social science theory and research, the success of other programs, and/or the program designers' own experiences and ideas.

Intervention or Treatment

Intervention or treatment comprises any activity or activities of a program that aims directly at changing a determinant. Intervention/treatment is, in other words, the agent of change within the program. The vital assumption made in the intervention/treatment domain is that by implementing these activities, the program changes the determinant and ultimately reaches its goals. For example, a treatment program for juvenile delinquency chooses to focus on a community's lack of accessible positive role models for youth. The intervention or treatment provided by the program is to team each youth with a volunteer, an accomplished professional or businessperson from the area, to serve as a role model. Volunteers are expected to spend 2 hours each week with the participant, providing guidance and encouragement related to school, home, and neighborhood. Once a month, the pair is asked to attend a community event or visit with a private or public organization. As the pair's relationship deepens, the program designers assume, the status of the volunteer and his or her personal interest in the youth will motivate the youth to identify with the volunteer and emulate his or her agenda of productive and beneficent activities. This will lower the odds of future involvement in delinquency. In many cases, an intervention or treatment has a number of elements. For example, alcohol abuse treatment is likely to include detoxification, individual and group counseling, and family therapy.

Some intervention programs can attain program goals without mediating by a determinant. Food relief programs in a disaster or warring region are a good example. A food relief program is regarded as successful as long as foods are distributed to and consumed by refugees. However, the great majority of intervention programs aim at changing knowledge, beliefs, behaviors, and/or skills. These kinds of programs usually require the intervention to change some determinants in order to affect goals or outcomes.

The terms *treatment* and *intervention* have been used interchangeably in the program evaluation literature. However, for health-related programs, at least, there is a subtle difference between the two concepts. In health-related programs, *treatment* is equal to caring for and, ideally, curing people who currently have some illness. *Intervention* more often refers to an effort to alleviate an existing problem, to ward off a potential problem, or to improve some aspect of quality of life. An intervention might sometimes comprise treatment. The evaluation principles and strategies discussed in this book can be applied to either treatment or intervention programs. For simplicity's sake, in the remainder of the book, the term *intervention* will be used, covering both meanings.

COMPONENTS OF THE ACTION MODEL

An action model is a systematic plan for arranging staff, resources, settings, and support organizations in order to reach a target population and deliver intervention services. This programmatic model specifies the major activities a program needs to carry out: ensuring an environment for the program that is supportive (or at least not hostile), recruiting and enrolling appropriate target group members to receive the intervention, hiring and training program staff, structuring modes of service delivery, designing an organization to coordinate efforts, and so on. It is vital to recognize that the impact made by a program's change model results jointly from the intervention's effect *and* the particulars of the program's implementation. The success of a job training program, for example, is not totally determined by its curriculum but also by the quality of its teachers, the motivation and attitude of its participants, the job search strategies employed, and the vigor of the local economy. The following discussion touches on all major elements—that is, the complete form—of the action model; it provides an exhaustive list, which may be much more than the evaluator requires in actual practice. (A rule of thumb is that large-scale programs may need all six elements, whereas small-scale programs may be just as effective with only a few of them.) Nevertheless, familiarity with the complete action model enables the evaluation practitioner to discuss more than one version of program theory. Access to the complete action model also helps in determining which components are important in a unique set of circumstances and in understanding how to simplify or otherwise modify the model to fit particular evaluation needs. The elements of the action model are the implementing organization, program implementers, associate organizations/ community partners, context/environment, target population, and intervention and service delivery protocols. From this list of elements, program evaluators can draw ideas about areas of potential focus within evaluations they are designing.

Intervention and Service Delivery Protocols

The change model for a program reflects general and abstract ideas about intervention that must be translated into the set of concrete, organized, implementable activities constituting its programmatic model. Basically, there are two requirements for this translation: an intervention protocol and a service delivery protocol. The *intervention protocol* is a curriculum or prospectus stating the exact nature, content, and activities of an intervention—in other words, the details of its orienting perspective and its operating procedures. To begin to ascertain the intervention protocol of a family counseling program, for example, answers to the following general questions are needed: What is the nature of the counseling? What is the content of the counseling? What is the schedule for the counseling? Specific answers to these might be generated by asking questions such as the following: Is the counseling based on behavior therapy? On reality therapy? On another kind of therapy? Will counselors proceed by following standardized documents? How many counseling sessions are planned, and how long will each last?

In contrast, the *service delivery protocol* refers to the particular steps to be taken in order to deliver the intervention in the field. The service delivery protocol has four concerns: client processing procedures, or how clients move from intake to screening to assessment to service delivery; division of labor in service delivery, or who is responsible for doing what; settings, which may be formal (e.g., at a program's office) and/or informal (e.g., in a client's home); and communication channels (face-to-face, telephone, mail, etc.). As an example, the service delivery protocol of a program addressing child abuse would provide answers to the following questions: Where will counseling take place—in a counselor's office or in clients' homes? Will each parent be counseled separately, or will they meet with the counselor together? At what point, if any, will child and parents be counseled together? In general, one place to look for the level of quality of a program is in its establishment (or lack of establishment) of an appropriate intervention protocol and service delivery protocol.

Implementing Organizations: Assess, Enhance, and Ensure Its Capabilities

A program relies on an organization or organizations to allocate resources; coordinate activities; and recruit, train, and supervise implementers and other staff. How well a program is implemented may be related to how well the organization is structured. Initially, it is important to ensure that the implementing organization has the capacity to implement the program, and strategies exist that can be helpful in determining this. For example, if a funding agency gets to choose the implementing organization from among several qualified candidates, that agency may be well equipped to determine which organization is most capable of implementing the program. In reality, however, such a pool of capable organizations ready for action may be missing. This is especially true for community-based organizations. Usually, an implementing organization's capacity to conduct the program must be built up. *Capacity building* involves activities such as training, technology transfer, and providing—financially and otherwise—for the hiring of experts or consultants to help plan and conduct the implementation.

Program Implementers: Recruit, Train, and Maintain Both Competency and Commitment

Program implementers are the people responsible for delivering services to clients: counselors, case managers, outreach workers, schoolteachers, health experts, and social workers. The implementers' qualifications and competency, commitment, enthusiasm, and other attributes can directly affect the quality of service delivery. The implementers' competency and commitment also have a direct effect on the quality of the intervention delivered to clients, and thus the effectiveness of the program in large part depends on them. Under the action model, it is important for a program to have a plan for ensuring competency and commitment among program implementers, using strategies such as training, communication, and performance monitoring/feedback.

Associate Organizations/Community Partners: Establish Collaborations

Programs often may benefit from, or even require, cooperation or collaboration between their implementing organizations and other organizations. If linkage or partnership with these useful groups is not properly established, implementation of such programs may be hindered. In the example of the spouse abuse treatment program, program implementers need to work closely with the court to develop procedures requiring convicted abusers to participate in treatment as part of their sentences. This program would meet with serious difficulty if it lacked a working relationship with the court or failed to win the support of judges. Under the action model, it is important to create feasible strategies for establishing and fostering relationships with associate organizations and community partners. As will be detailed in Chapter 5, this element is most important when an evaluator is asked to take a holistic approach to help program designers and other stakeholders plan and develop a program.

Ecological Context: Seek Its Support

Ecological context is the portion of the environment that directly interacts with the program. Some programs have a special need for *contextual support,* meaning the involvement of a supportive environment in the program's work. (Indeed, most programs can be facilitated to a degree by an environment that supports the intervention processes.) A program to rehabilitate at-risk juveniles, for instance, is more likely to work when it obtains the support and participation of juveniles' families and friends. Both *micro-level contextual support* and *macro-level contextual support* can be crucial to a program's success. Micro-level contextual support comprises social, psychological, and material supports that clients need in order to allow their continued participation in intervention programs. For example, under current welfare reform laws, in order to receive benefits, mothers must attend job training or find work. But these reforms present two immediate problems: Is transportation available to get the women to the workplace? And who will care for the children while they work? A welfare-to-work program is hardly manageable without tackling these issues. Furthermore, clients may be more likely to participate seriously in programs when they receive encouragement and support from their immediate social units (typically family, peer group, and neighborhood). When program designers or implementers realize that micro-level contextual support could play an important role in an intervention, it is up to them to try to build this support into a program's structure. For example, designers of an alcohol abuse program might organize a support group for clients that includes family members and peers who encourage and support them during and/or after intervention. In addition to micro-level contextual support, program designers should consider the macro-level context of a program; that is, its community norms, cultures, and political and economic processes. These, too, have the ability to facilitate a program's success. A residential program for the mentally ill can anticipate real difficulties if the local community has a generally hostile attitude toward its clients. But if an adequate campaign for community support of such patients is one component of the residential program's implementation, these difficulties may be alleviated. In any case in which stakeholders believe macro-level contextual support to be crucial to their program's success, the generation of this kind of support should be included as an element of their program.

Great effort goes into ensuring the capabilities of implementing organizations, establishing collaborations with associate organizations, and winning contextual support, if these are truly done well. Finding resources with which to make the effort can be a challenge. There is a worthwhile payoff, however. If a program does succeed in these activities, it is considered an *ecological,* or *multilevel, intervention program:* that is, a program with goals not just for individual clients but also for the surrounding community. Ecological programs may be likelier to attain their goals than are programs concentrating simply on client issues. This element signals a need to take a holistic approach to conduct program evaluation.

Target Population: Identify, Recruit, Screen, Serve

The target population or group is the people that the program is intended to serve. Concerning target population, three assumptions that often figure in evaluation are the presence of validly established eligibility criteria, the feasibility of reaching eligible people and effectively serving them, and the willingness of potential clients to become committed to or cooperative with (or at least agreeable to joining) the program. Faced with resource constraints, a program usually cannot provide services to everyone in a target population. Therefore, it needs a clear and concrete boundary for eligibility. Criteria must also be established by which the program determines which specific populations it will recruit. For example, the target population of one Head Start program is preschool children from disadvantaged families residing in a particular community. Similarly, an HIV prevention program in one community chooses to serve addicts who inject drugs rather than trying to target the entire high-risk population. A program is usually regarded as ineffective if it finds itself serving the wrong population or failing to reach enough members of the right population. A nursing care program intended to serve low-income elderly people, for example, has failed if its services benefit many comparatively well-to-do people. Similarly, a job training program that is well funded and well run will have failed if it produces only a handful of "graduates."

Whether or not clients are prepared to accept intervention also can affect program outcomes. Especially for labor-intensive types of programs, client screening and assessment are vital. Identification of *actual* needs is vital, and information from assessment can suggest whether a client needs services in addition to the central intervention. For example, when assessment reveals the need, clients can be referred by program staff for housing assistance, mental health care, education, employment, or other social services. A labor-intensive program must be certain of its clients' readiness for intervention; *client readiness* being the extent to which an individual's mental and physical state permits his or her acceptance of an intervention. If clients are not mentally and physically ready for it, intervention is unlikely to work. *Mental readiness* of a client is the degree of his or her willingness to recognize a problem or deficiency, or the degree of motivation to accept an intervention. For example, a person who insists alcohol is not a problem for him or her will probably not succeed in an alcohol abuse counseling program. Clients also exhibit varying degrees of *physical readiness* for interventions. Health status affects delivery of some interventions. For example, counseling clients about HIV prevention can be difficult when they suffer from severe mental health problems or have no food or shelter. In such a case, the successful intervention program is likely to provide case management or referral services to meet basic needs prior to beginning intervention. Similarly, a client still under the influence of alcohol is no more physically ready than mentally ready for intervention. Trying to deliver alcohol counseling services is futile until the client has completed a detox program; alcohol abuse intervention starts once the client is sober.

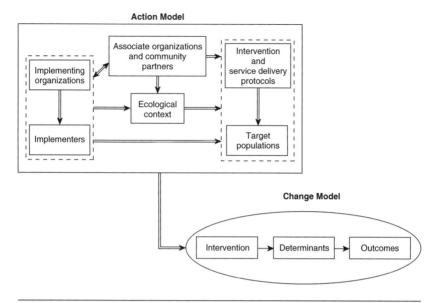

Figure 2.1   Conceptual Framework of Program Theory (Basic Form)

RELATIONSHIPS AMONG COMPONENTS OF THE CONCEPTUAL FRAMEWORK OF PROGRAM THEORY

It is important to understand relationships among program components. The relationships among components can be pictured as in Figure 2.1.

In general, program components need to be organized or connected in a meaningful way in order to achieve the goals. Figure 2.1 shows how an action model is implemented in order for a change model to activate the causal process. The double-banded arrows between two components within the action model represent a sequential order between these two components. That means that the completion of some component provides the basis for completing the next component. For example, in the figure, the double-banded arrow from "implementing organization" to "implementers" indicates that it is usually a requirement to have a capable implementing organization in place in order to adequately recruit and train implementers. With a spouse abuse intervention program—or virtually any program—this means that there must be an organization responsible for implementing the program before counselors or clients can be recruited. In other words, the relationships among components of the action model represent a kind of "task order" relationship: Some components must be in place and complete before others can be brought in line. The only exception is the two-way double-banded arrow between the box of the implementing organizations and the box of associate organizations and community partners. This means very often the associate organizations and community partners collaborate with the implementing organizations in planning program activities at the same time.

On the other hand, the solid arrows within a change model in Figure 2.1 depict causal relationships in the change model. A causal relationship exists between elements when changing one creates change in the other(s). A solid arrow leading from an intervention to a determinant represents the model's assumption of a causal relationship between the two. In the spouse abuse program, the model assumes that group counseling has the power to create anger management skills in abusers and to teach them about the legal punishments associated with spouse abuse.

The conceptual framework should make clear that the action model must be implemented appropriately in order to activate the transformation process in the change model. In order for a program to be effective, its action model must be sound and its change model plausible; its implementation is then also likely to be effective. For example, in order for an HIV prevention outreach program to succeed, it needs to coordinate activities, reach the target group, and provide the group with adequate exposure to the prevention message; it must also determine which activities will strengthen the target group's knowledge of risk prevention, which should manifest itself in decreased high-risk sexual behavior. This conceptual framework of program theory should be useful to evaluators charged with designing an evaluation that produces accurate information about the dynamics leading to program success or program failure.

If evaluators and stakeholders want mainly to highlight the relationships among the components of program theory, Figure 2.1 is sufficient. However, Figure 2.1 does not address the relationships among program, environment, and feedback discussed in Chapter 1. For evaluators and stakeholders interested in elaborating these further relationships, a comprehensive diagram, such as Figure 2.2, is necessary.

In Figure 2.2, the large square around the program represents its boundary. Everything within the large square is part of the program; all that is outside the square is "environment," providing the program with necessary resources and support (in other words, its inputs), or, perhaps, working against implementation of the program. Figure 2.2 shows that, generally, a program starts with the acquisition of resources from the environment and the development of an action model. Fueled by the acquired resources, the action model can be

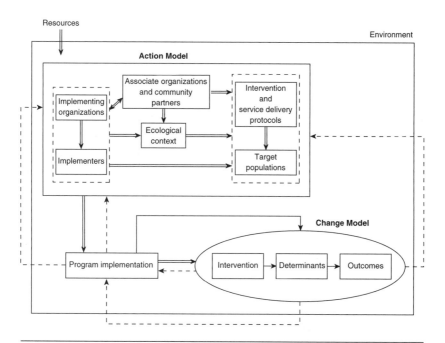

Figure 2.2 Conceptual Framework of Program Theory (Comprehensive Form)

implemented in order to activate the change model. It is the operation of the change model that leads to the attainment of program goals. Solid arrows joining an action model to a change model indicate that, strictly speaking, whatever effect the program has on the outcomes is not due to the implementation of intervention alone but to a joint effect of the implementation of intervention and the implementation of other factors in the action model. Evaluation feedbacks are represented in dotted arrows. The evaluation feedback in the figure comprises information about how the action model was implemented in the field, such as whether the program reached the right target population.

Similarly, the dotted arrow from the implementation to action model indicates that evaluation feedbacks from the implementation can be used to improve the *planning* or the development of the action model. The dotted arrows from the change model to the implementation and action model indicates that the information from the causal process of the change model can be used to improve or modify the implementation process or the planning of the action model. The conceptual framework provides two distinct general evaluation feedbacks: the internal and the external. The dotted lines in Figure 2.2 represent evaluation feedback and feature two sets of "feedback loops." Each set of evaluation feedback loops indicates one path that program evaluation can follow to obtain information vital to program improvement. Each path accommodates distinct audiences and purposes. Therefore, the evaluation approaches and strategies involved in various evaluation feedback loops can be quite dissimilar. The first set, constrained inside the program boundary, is for an internal audience of program implementers, administrators, and others who deal with programmatic concerns and service delivery matters on a daily basis. This audience wants from the program evaluator timely information on whether a program is operating smoothly in the expected manner. If there are difficulties, the internal audience wants to understand, if possible, the sources of the problems as well as the likely remedies. This aspect of evaluation is called *internal use evaluation* or *development-oriented evaluation*. Strategies and techniques of internal-use evaluation must be flexible and creative, and they must be accomplishable quickly. If a program is not on the right track, its course must be corrected before too much time and energy are wasted.

The remaining set of feedback loops in Figure 2.2 passes to the environment and then back again to the program. This is the *external feedback loop,* incorporating both scrutiny by the environment and improvements from the program itself. Conducting external feedback evaluation requires more resources and more time than conducting internal feedback evaluation. The audience for external feedback is funding agencies; decision makers; interested groups; the public at large; and the stakeholders who work inside the program, such as program directors and implementers. The external feedback loop represents a mechanism that delivers to the environment information about the merits of a program, what changes the program may need, and the appropriate general direction for the program in the future. There are two types of evaluation relating to the external feedback loop. One is intended to serve accountability needs and is called *assessment-oriented evaluation.* The other is designed to serve both accountability and program improvement needs and is called *enlightenment-oriented evaluation.*

These different types of evaluation will be discussed in detail in the remainder of the book.

APPLIED PROGRAM THEORY: AN EXAMPLE

A good example for the application of program theory for program evaluation is found in an evaluation of an antismoking program (Chen, Quane, & Garland, 1988). Program designers devised a comic book with an antismoking

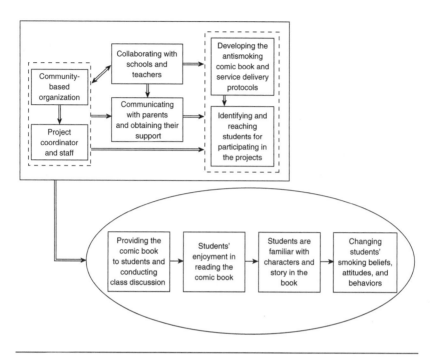

Figure 2.3　　　Program Theory of an Antismoking Campaign

SOURCE: Adapted from Chen et al. (1988).

story as an intervention to change students' knowledge, attitudes, and behavior concerning smoking. Program designers expressed a desire for an outcome evaluation of the program that would provide information needed to make improvements to the program. The program theory supporting the program was stakeholder theory, stemming from the program designers' own ideas and experiences. Evaluators conducted intensive interviews to clarify the stakeholder theory.[1] The stakeholder theory is illustrated in Figure 2.3.

Change Model

The program designers' main idea for the program came from their observation that teenagers are fond of reading and collecting comic books. Accordingly, they thought a comic book that conveyed an antismoking message would create an opportunity for students to absorb selected facts and change their attitudes and behavior concerning smoking. More specifically, the program designers' change model contained two determinants in a sequential order: the students' enjoyment of reading comics and students' familiarity with the characters (heroes and villains) and story. The designers hypothesized that these determinants would lead to stronger antismoking beliefs and behaviors.

Action Model

The program designers had in mind a story, characters, and even scripts, and they collaborated with a community-based organization to implement the project. They proposed hiring a comic book artist to draw the pictures and a project coordinator and staff to run the program. They named a target population— young people attending middle school—and sought support from principals, teachers, and parents in encouraging students to participate. They planned to distribute the comic book in health classes.

After the evaluation was conducted, results showed the program to be well-implemented based on the proposed action model. Results for the change model were mixed. Although students read and kept the comic book, possessing it as expected, these determinants alone were not sufficient to translate into attainment of the intervention goals. The evaluation showed where the program had misstepped in the change model. The information was useful for program stakeholders to design a better program in the future.

RELATIONSHIPS BETWEEN "PROGRAM THEORY" AND "LOGIC MODEL"

The terms *program theory* and *logic model* have been used interchangeably in the literature, but, in fact, the two serve separate purposes. A logic model is a graphical representation of the relationship between a program's day-to-day activities and its outcomes. It does resemble program theory in its employment of diagrams to aid the analysis of programs, and yet conceptualization of programs within program theory and within the logic model is utterly distinct. Program theory is a systematic configuration of prescriptive and descriptive assumptions underlying a program, whereas the logic model stresses milestones like components. Wholey (1979) rendered the logic model in two primary parts: the program components, and the goals and effects of the program. *Program components* are activities that can, whether conceptually or administratively, be grouped together.

Building on Wholey's work, subsequent versions of the logic model have tended to add parts to the logic model. One popular twist on the model is the version developed by the United Way of America (1996). With it, evaluators

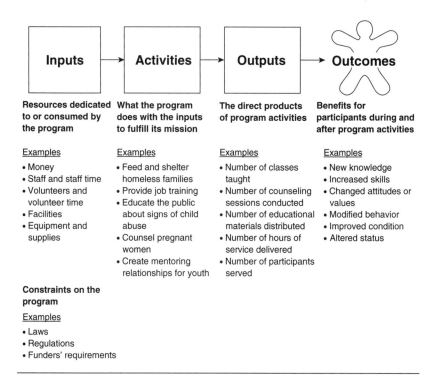

Figure 2.4   The United Way's Logic Model

SOURCE: Reprinted with permission of the United Way.

of United Way programs consistently examine inputs, activities, outputs, and outcomes. In this particular logic model, *inputs* are defined as resources dedicated to or consumed by the program: money, supplies, staff, and even ideas. *Activities* in this logic model comprise actual services or work that the program provides to fulfill its mission. Examples include recruiting and training staff, counseling clients, providing referral services, and educating the public. In the United Way model, *outputs* are defined as the direct products of program activities: number of clients served, number of classes taught, amount of goods distributed, and so on. Finally, this logic model defines *outcomes* as the actual benefits resulting from program activities. Examples are improved health, new knowledge, better skills, and higher income. Relationships of these elements to each other are illustrated in Figure 2.4.

It is possible to view the logic model as a simple version of program theory. The logic model's parts essentially comprise generic categories of program theory elements. For example, under the logic model's heading "inputs," program theory's *resources* as well as its entire *action model,* with all its components, could easily be subsumed. However, the logic model and program theory serve distinct purposes. The logic model's largest concern is how to develop an *evaluable* program, conduct evaluability assessments, and monitor a program's performance. Every component of the logic model can be translated into indicators or measures of performance. For example, consider a logic model of a school-based dental care program. The logic model would focus on quantification of the program's activities, such as the number of students participating, the number of dental health brochures distributed, the number of service and education sessions conducted, and the number of schools participating. To that end, it provides milestones for measuring a program's progress along the way (a topic discussed extensively in Chapter 8).

Program theory, moving beyond milestones and quantification, would strive to ask what makes a program *sound,* and how this is accomplished. For example, how does reading a brochure initiate better dental care? At the same time, although program theory aims at devising comprehensive assessment strategies for the evaluation of programs (refer to Chapters 4, 5, 7, and 10), many of the principles and techniques of program theory that this book discusses can also be applied within the logic model.

SOME ADVANTAGES OF USING THE PROGRAM THEORY CONCEPTUAL FRAMEWORK

Holistic Assessment

When designing and conducting an evaluation, using the conceptual framework of program theory to develop contingency principles offers several advantages. First, the conceptual framework allows a holistic approach to evaluating the merits of a program. Following the conceptual framework, an evaluation can explain how and why a program achieves a particular result by illustrating its means of implementation as well as underlying mechanisms that influence it. Let us look at another example: A new curriculum has been introduced in a school in the hope of raising students' test scores. By proceeding from the conceptual framework, the evaluation of the new curriculum will do three important things: obtain information about achievement of goals, ask how effectively the action model was implemented, and explore the role of any underlying causal mechanisms. Keeping the conceptual framework in mind, the evaluator will be prompted to document the curriculum's implementation, how the students were recruited, and how the teachers taught the curriculum and were motivated to use it. The conceptual framework also prompts queries about underlying causal mechanisms: Are achieved goals truly attributable to innovations in the curriculum? Or have goals been reached by "teaching the test" to students, or taking a punitive approach to low scorers? The conceptual framework addresses issues in both the action model and the change model, so it helps the evaluator achieve a balanced view from which to assess the worth of a program. This kind of assessment of what works and what does not work prevents throwing the baby out with the bath water.

Comprehensiveness of Information Needed to Improve Programs

An evaluation that examines how a program's structure, implementation procedures, and causal mechanisms actually work in the field will provide information that can be very useful in program improvement. For example, if the government of a developing country wants to offer low-interest loans to farmers for costly machinery or fertilizers, program evaluators can work with program designers and other key stakeholders to critique assumptions that underlie their work. For example, will farmers be well informed about the terms of the loans? Will local loan officers welcome a new loan program and do their part to solicit and approve loan applications under it? Evaluation of a program's underlying assumptions brings to light information that helps key stakeholders see why the program is likely to work (or did work) well or is not likely to work.

Delineation of a Strategy to
Consider Stakeholders' Views and Interests

Evaluators such as Patton (1997) and Fetterman, Kaftarian, and Wandersman (1996) have argued forcefully that the design of an evaluation suffers without adequate input from stakeholders. Earlier in this section, it was noted that the conceptual framework requires evaluators to be familiar with stakeholders' assumptions about their program theory. (These assumptions can be science-based or based on personal beliefs and experiences.) When stakeholders' interests and views are given due consideration during the designing process, the evaluation's relevancy and usefulness burgeon.

Flexibility of the Research Methods
Harnessed to Serve Evaluation Needs

In taking a contingency approach toward research methods, the program theory conceptual framework offers a guide to the flexible application of methods in order to best address particular evaluation issues. Few programs can be truly called identical. They all vary in structure, processes, maturity, environment, and stakeholder needs. Research methods should be tailored to meet evaluation needs, *not* vice versa. By taking on conceptual issues, the evaluator is liberated from the rigidity—the dogmatism, even—of method-driven evaluation and its ironclad research methods.

Aid for Selecting the Most Suitable Approaches or Methods

Intense conflicts among evaluators have existed over such major theoretical issues as the nature of evaluation and the chief end of evaluation, and also over pragmatic matters such as the best methods available to our field. These persisting conflicts may have created confusion. But by taking a contingency approach, the program theory conceptual framework ensures that the merits of a principle, strategy, or method are judged individually and in context, rather than absolutely. By thinking contextually, each evaluation principle or method is allowed its distinct value, apart from the values inherent in competing principles, and evaluators are freed to weigh them all. In this way, it is the confusion, not the options, that is minimized. The framework helps make the options manageable by identifying those circumstances under which certain concepts and techniques are most appropriate. The conceptual framework, then, has at its heart the importance of situational factors for evaluation. This contingency view has the potential to narrow the gap between evaluation theory and evaluators' practice.

SCIENTIFIC THEORY VERSUS STAKEHOLDER-IMPLICIT THEORY AS A BASIS FOR PROGRAMS

In developing a program, the adopted program theory—and particularly the change model—can be based either on well-defined scientific theory or on stakeholder-implicit theory. The characteristics and merits of the two options require some discussion.

Programs Based on Scientific Theory

When scientific theory or evidence is relied on for the development of a program, it is a *scientific theory-based program*—and it may be bursting with fertile information about which determinants will make the program work. This is especially so of scientific theory-based programs in the areas of health promotion and disease prevention. The science available in these areas helps program designers and evaluators understand why a particular determinant should have the power to shape outcomes (Bartholomew et al., 2001; Witte, Meyer, & Martell, 2001). Scientific theories are usually well tested, so choosing to design programs upon their principles can eliminate trial-and-error searching for determinants. In fact, scientific theory-based programs are often developed with scientists and scholarly researchers at the helm.

For example, in order for mothers to reduce passive smoking by infants, Strecher et al. (1989) applied Bandura's (1977) social learning theory to identify determinants and design an intervention. According to social learning theory, behavioral change (and the maintenance of it) arises from new expectations concerning outcomes of a behavior. This is called *outcome expectation.* Social learning theory is also concerned with *efficacy expectation,* which comprises beliefs concerning one's own capability to engage in the particular behavior. The home-based program was targeted at mothers and was designed with their outcome expectations *and* efficacy expectations in mind. In applying social learning theory to infants' passive smoking, the outcome expectation became mothers' perceptions of what happens to an infant exposed to environmental tobacco smoke. The efficacy expectation became mothers' perceptions about their own ability to create and maintain a smoke-free environment for their infants. Strecher et al. included outcome and efficacy expectations as two determinants of the intervention. Figure 2.5 illustrates the research team's scientific theory-based program.

There are clear advantages to basing a program design on scientific theory. The program will obtain respect more readily. In addition, because scientists often take the lead in such programs, ideal testing conditions may be available for evaluations; this increases the chance of finding the thing that will make a program tick. But danger exists with science-based programs as well. They tend to focus heavily on academic interests rather than stakeholder interests. Stakeholders often find that programs tied to scientific theory are too controlled and fail to sufficiently reflect the real world. This issue is discussed intensively in Chapter 9.

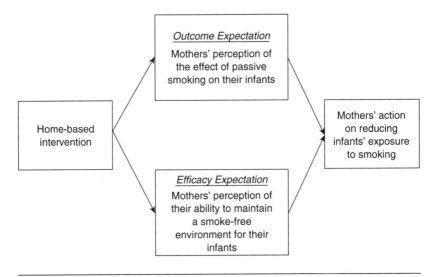

Figure 2.5   A Scientific Theory-Based Program to Reduce Infants' Passive
                Smoking

SOURCE: Adapted from Strecher et al. (1989).

The program theory conceptual framework is of value to those needing
to assess scientific theory-based programs for two basic reasons that are
discussed thoroughly in Chapters 9 and 10. First, the framework fosters,
in assessments, the integration of those causal mechanisms identified by
the pertinent scientific theory or theories. Too often, evaluations of scientific
theory-based programs seem limited to simply gauging the effects of intervention on outcomes; virtually ignored are ideas about underlying causal
processes, although these are what theories are largely about. Using the program theory conceptual framework fends off the temptation to conduct "black
box" evaluation[2] and assure that the scientific theory underlying the program
is evaluated. Second—and on the other hand—many science-based programs
under development could stand to step away from the ideally controlled laboratory situation and toward more lifelike scenarios in order to increase their
relevancy to practice. The conceptual framework of program theory can assist
scientists in designing assessments that more aptly mimic the real world.

Programs Based on Stakeholder Theory

A program founded mainly on stakeholder-implicit theory is a *stakeholder theory-based program*. The majority of intervention programs that

operate in a community have not been designed by scientists but rather by stakeholders such as program designers, program directors, and program staff. Each of these individuals perceives the nature of the problem in an individual way and develops a personal preference for a particular solution to the problems confronted. These perceptions and preferences may come from past experiences, conventional wisdom, discussion with peers, advice from experts, scientific theories, acquaintance with similar programs—even hunches. In other words, stakeholders have their own program theory. Of course, stakeholder theory is not usually rendered as an explicit and systematic statement the same way as scientific theory does. Stakeholder theory is implicit theory. It is not endowed with prestige and attention as is scientific theory; it is, however, very important from a practical standpoint because stakeholders draw on it when contemplating their program's organization, intervention procedures, and client-targeting strategies. Stakeholders' implicit theories are not likely to be systematically and explicitly articulated, and so it is up to evaluators to help stakeholders elaborate their ideas.

An illustrative case is Clapp and Early's (1999) study of alcohol and drug abuse prevention programs for young Hispanic students. These programs tended to be vague as to their components and rationales. Clapp and Early met with focus groups of clients, staff, and parents from each program in order to clarify the stakeholders' implicit theories. In one school-based program, the researchers ironed out two intervention elements desired by stakeholders: English language acquisition and a course in coping with feelings. The stakeholder theory assumed that these two elements would sequentially affect the two program determinants: identifying with the larger culture and internalizing that culture's norms. When fully exploited, the stakeholders theorized, the two determinants would produce an outcome of reduced alcohol and drug abuse. The change model implicit in the stakeholders' beliefs is delineated in Figure 2.6.

The conceptual framework of program theory can help evaluators to "read" stakeholders' implicit theories in three ways. The first involves making the implicit explicit as well as systematic. If asked to participate from the very beginning of the program, an evaluator working from the conceptual framework sees places where the program would benefit from spelling out the stakeholders' assumptions. The conceptual framework also suggests which assumptions might be weak or which elements of a theory are missing (see Chapters 4 and 5 for further discussion).

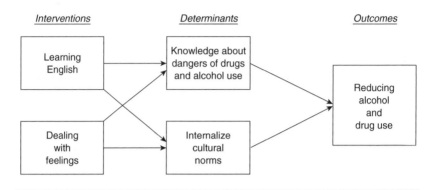

Figure 2.6   A Stakeholder Theory-Based Alcohol and Drug Abuse Prevention
                    Program for Hispanic Students

SOURCE: Adapted from Clapp and Early (1999).

The second way the framework helps is by saving time in situations in which quick feedback could confirm that a program design is on track. With timely feedback, a program or implementation can be developed to a level at which it will operate smoothly and functionally with little waste. The conceptual framework offers a direction for evaluation design that, although ready-made, is also amenable to tailoring, a point elaborated in Chapters 5 and 6 of this book. Third and last, the conceptual framework's versatility allows evaluators to assess comprehensively the overall quality of implementation and program effectiveness, and at the same time identify strengths and weaknesses in its elements (see Chapters 7 and 10).

APPLICATIONS OF THE
FRAMEWORK PRESENTED IN THE BOOK

As the reader progresses further into this book, the three general purposes of the program theory conceptual framework are explained. One purpose is to become the underpinning of a taxonomy of evaluation circumstances and accompanying evaluation strategies and approaches. Contingency principles inherent in this taxonomy are discussed and show how evaluators and stake-holders can set about meeting their evaluation needs. Chapter 3, especially, elaborates the discussion of this particular purpose of the framework. A second purpose of the framework is to lay out for the practitioner those evaluation approaches best suited to the *program planning, implementation,* and *outcome* stages, suggesting some applications for these approaches. (A catalog of evaluation activities is dispersed throughout Chapters 4 to 10.) A final purpose of the conceptual framework is to introduce theory-driven evaluation as one option open to evaluators. Theory-driven evaluations are discussed in Chapters 7 and 10.

NOTES

1. How to clarify stakeholders' program theory will be discussed intensively in Chapters 4 and 5.

2. A black box evaluation is an evaluation that mainly assesses the relationship between intervention and outcome. A program may be based on a scientific theory. However, it is possible that an evaluation on the program is a black box evaluation. This issue will be further discussed in Chapter 10.

THREE

A PRACTICAL
EVALUATION TAXONOMY

Selecting the Evaluation
Approach That Works

———•◦•———

I t is vital that practicing program evaluators know which evaluation strategy and which evaluation approach, out of the many available, will be best suited to meet stakeholders' fluctuating needs. The "fish story" that opens the next section illustrates the value of this knowledge. The story is followed later in the chapter by a *taxonomy* of practical program evaluation means and ends: comprehensive, systematic guidance to use while weighing the circumstances and needs of one's evaluation assignment against the strengths and short-comings of various evaluation strategies and approaches.

THE ART OF FISHING?
THE ART OF PROGRAM EVALUATION?

Whereas discussions of individual evaluation approaches and methods are encountered frequently in the literature (e.g., Mark, Henry, & Julnes, 2000; Rossi, Lipsey, & Freeman, 2004; Shadish, Cook, & Leviton, 1991), there is much less written about issues that affect the profitable selection of one approach or strategy from among the many. The information that is available tends not to be systematically presented. An analogy of fishing suggests how this gap in program evaluation can make life more difficult for the program evaluator. To go fishing one needs, first of all, equipment—the poles and lines, hooks, sinkers, floaters, bait. Without this basic equipment, fishing is (for most humans) impossible. Possessing equipment and knowing how to employ it, however, do not guarantee success. Choosing the wrong equipment from all that is available—the wrong size fishing line, or the wrong hook, or an inappropriate bait for a given fishing spot—probably means turning up empty-handed, even if one handles that line, hook, or bait magnificently.

The vital, and yet limited, role of equipment in the art of fishing is seen clearly in fly fishing, the technique in which the angler continuously casts and retrieves a line tipped with, or baited with, an artificial fly. With tackle and casting know-how, any person can go through the motions of fly fishing. Those who can be counted on to catch fish, however, are those accomplished in the art. A good fly-fishing angler knows how to choose the right place and time as well as the right artificial fly. Fly fishing masters have learned to habitually consider such things as season, currents, play of light and shade, and types of surrounding vegetation, in addition to their equipment. These masters can select just the right fly to mimic whatever real fly would inhabit a given area at a given moment, so that the fish pursue the fly without suspicion.

Productive fishing is more than equipment and the ability to operate it. In order to catch fish, it helps to know fish habits and habitat: favorite foods, favorite pools or banks, responses to weather, and so on. Upon this kind of understanding is based "the art of fishing." Mastery of this art may result from trial and error over a long period of time or, more efficiently, from instruction by someone experienced in fishing. It may also be developed by studying authoritative books.

What can program evaluators learn from the art of fishing? To begin with, consider evaluation approaches and methods as being analogous to fishing equipment. Only when we have our evaluation approaches and methods down pat can we set about catching a fish. But although this is a necessary condition, it remains an insufficient one. Like an average fishing enthusiast with a tackle box and some manufacturer's instructions, an evaluator familiar with evaluation approaches and methods can try his or her luck. But there is no more guarantee of a decent evaluation than of landing a catch. Like fishing the wrong spot or using the wrong bait, missing important issues in evaluation design finds the adventurer returning home with no prize. Evaluation becomes productive only when we go beyond methodology to ply the waters with theoretical and also with contextual knowledge.

THE EVALUATION
TAXONOMY AS A SOURCE FOR
THE ART ASPECT OF PROGRAM EVALUATION

Practical evaluators must be able to apply evaluation approaches correctly— the scientific aspect of program evaluation. But before they can do so, they must be able to select the evaluation approach that complements the needs and realities they face—the art aspect of program evaluation. The evaluation taxonomy around which this book is organized is offered as a means of enhancing understanding of evaluation's art aspect. A taxonomy provides any discipline with an important tool for classifying phenomena, guiding the selection of research strategy, facilitating communication, and developing knowledge. In program evaluation, a taxonomy can help with conceptualization of evaluation needs, focusing of evaluation activities, and identification of available evaluation means that suit a program's needs and realities.

An early example of classifying evaluation activities is the distinction of *formative* from *summative evaluation* proposed by Scriven (1967). Formative evaluation is concerned with designing and using evaluation to improve a program, whereas summative evaluation is concerned with designing and using evaluation to judge a program's merit (Shadish et al., 1991). Chen (1996, 2004) proposed a *taxonomy of program evaluation,* one built around the program stage that is the desired focus of the evaluation, as well as around the desired function of the evaluation (either improvement or assessment). This section of the book furthers this earlier work on the notion of a taxonomy. It introduces a systematic taxonomy for daily use by program evaluators in the field. This practical taxonomy of program evaluation ends and means is represented in Table 3.1.

At left in the table are several "classes" of potential evaluation needs of stakeholders at each stage of program growth. These classes are linked to numerous evaluation approaches and strategies on the right. The remainder of this chapter fills out the table, with discussion of stakeholders' needs and subsequent discussion of evaluation approaches and strategies available to meet them.

Table 3.1   A Practical Taxonomy for Program Evaluation Means and Ends

Program Stages and Evaluation Purposes	Evaluation Strategies	Evaluation Approaches
Program Planning Stage		
Providing pertinent information and assistance to help stakeholders in developing program rationale and plan	Background information provision	Needs assessment Formative research
	Development facilitation	Conceptualization facilitation Concept mapping
	Troubleshooting	Relevancy testing Pilot testing Commentary or advisory
	Development partnership	Bilateral empowerment evaluation
Initial Implementation Stage		
Providing timely information on implementation problems and the sources for assisting stakeholders to fix the problem and stabilize the program	Troubleshooting	Formative evaluation Program review/ development meeting
	Development partnership	Bilateral empowerment evaluation
Mature Implementation Stage		
Providing information on implementation problems and the sources for improving the implementation process	Troubleshooting Development facilitation	Formative evaluation Program review/ development meeting Conceptualization facilitation Concept mapping
Assessing the quality of implementation for serving accountability needs	Performance assessment	Fidelity evaluation
Monitoring progress of implementation	Performance monitoring	Process monitoring
Holistic assessment of implementation process	Enlightenment assessment	Theory-driven process evaluation
Outcome Stage		
Assessing whether program is ready for outcome evaluation	Development facilitation	Evaluability assessment
Monitoring client's progress of outcomes	Performance monitoring	Outcome monitoring
Assessing the effect on outcomes	Performance assessment	Efficacy evaluation Effectiveness evaluation
Holistic assessment of the program effect for serving accountability and program improvement needs	Enlightenment assessment	Theory-driven outcome evaluation

A PRACTICAL TAXONOMY OF PROGRAM EVALUATION MEANS AND ENDS

Classifying Stakeholder Needs at Particular Program Stages

The first row of Table 3.1 (illustrating the *practical taxonomy of program evaluation means and ends*) lays out four stages of a program's growth and the nature of each stage. Evaluation requirements associated with each stage are highlighted. Stakeholders' evaluation needs vary across the stages of program growth (Chen, 2004). Evaluators can best understand stakeholders' evaluation needs if the evaluators are provided with information on the stage(s) the stakeholders are interested in evaluating. There are four program stages in the taxonomy: planning, initial implementation, mature implementation, and outcome.

It is often assumed that a program will move sequentially through these stages. This makes sense, but, in reality, programs could go back and forth between stages in a nonlinear fashion. As stages skip around, evaluation needs do, too. As an example, consider a program in its mature implementation stage. The program has been troubled by several major problems with service delivery, and its stakeholders decide to revise the program plan and return to the initial implementation stage. Thus, their evaluation needs are different from what they once were. Similarly, if, in its outcome stage, a program is found to be ineffective, its stakeholders could move to redesign the program, returning to the initial planning stage.

Evaluators in the field are asked to conduct evaluations at any stage and for various combinations of stages. When the program is an established one, evaluation of its implementation and outcome stages is common; start-up programs, too, need evaluations at these stages. Start-up programs also frequently require evaluators during the planning stage and initial implementation stage. In the following paragraphs, the evaluation needs characteristic of each stage are discussed.

Program Planning Stage. The first of the four stages is the *program planning stage.* This is the very beginning. Stakeholders at this stage—for example, program designers—are developing a plan that will serve as a foundation for organizing and implementing a program at some future date. As we have seen, programs can be complex; stakeholders often seek considerable help from experts with the hope of winding up with a plan of truly good quality. Today, evaluators are often found among these experts. In the program planning stage, stakeholders' primary evaluation need is to learn from evaluators the evaluation concepts, strategies, and activities that can help in the design and development of a program rationale and a program plan.

Early in Part II of this book, we read that program evaluation has, across much of its history, focused on outcomes. Lessons from the field, however, have plainly taught that program failures are often essentially *implementation* failures, and evaluation focus has gradually broadened to include process evaluation. The current view is that a major part of implementation failure can be traced to poor program planning and development. Evaluators can make important contributions in these areas where attention is most needed.

Initial Implementation Stage. The second stage cited in Table 3.1 is the *initial implementation stage.* As a program plan begins to be put into action, much can go wrong. During the initial implementation stage, a program's course can be highly fluid and unstable. At this point in development, stakeholders' evaluation needs include timely feedback on the major implementation matters and also identification of the sources of problems. These kinds of data can help the stakeholders to troubleshoot the implementation problems and to quickly stabilize the program.

Mature Implementation Stage. The *mature implementation stage* follows the initial implementation stage at a point when implementation of the program has settled into routine activities. Rules and procedures for conducting program activities are now well established. Stakeholders are likely to be interested in one or more of the following: continued unearthing of the sources of immediate problems, generation of data reassuring to those to whom stakeholders are accountable, and program improvement. Even in maturity, a program is subject to problems such as clients' dissatisfaction with services. A wise course for stakeholders in a case like this is to seek timely information from evaluators on the cause of problems. Identifying and resolving problems are key to improving a program. Furthermore, as a program matures, stakeholders may think more about their accountability. Data illustrating the effectiveness of implementation, or the efficiency of service delivery, are useful to stakeholders, who often ask evaluators to find such data if they exist. Finally, within the mature implementation stage, stakeholders begin to look for strategies of improvement (tying in to their need to be accountable, perhaps). They call on evaluators to provide information from a holistic standpoint through process evaluation that goes beyond assessing the quality of implementation to strengthening the program processes.

Outcome Stage. The fourth stage of program growth is known as the *outcome stage.* Following a period of program maturity, stakeholders inside and outside the program want to know whether the program is achieving its goals. An evaluation at this point can serve any of five primary evaluation needs. First, stakeholders may rely on evaluators to determine if a program is ready for outcome evaluation. It may not be and, if it is not, evaluators may be asked for help in building the program's "evaluation capacity." Second, stakeholders may want to monitor their clients' progress. Third, stakeholders may ask for information on what the program would be achieving if it existed in the ideal environment. (Such information can also help stakeholders decide if a program should be expanded to other people or settings.) Fourth, stakeholders may seek to know in detail the program's effects in its real-world setting because these, obviously, have a direct bearing on practice. Finally, some stakeholders may ask evaluators to go beyond traditional evaluation and its single-minded focus on assessment. They may want an evaluation that serves both accountability and program improvement needs.

Dynamics of Transition Across Program Stages

Intervention programs are goal-oriented activities. Ideally, the program moves directionally through the following stages: program planning; initial implementation; mature implementation; and, finally, outcome. For the convenience of the reader, the remaining chapters of this book are arranged according to an ideal sequence of program stages. In reality, however, the program stages may not be linear at all. For example, due to a stakeholder's dissatisfaction with the direction of a program, or due to political pressure calling for a change, a program could move from the mature implementation stage back to the planning stage rather than moving forward to the outcome stage. Similarly, a program at the outcome stage may make an overhaul of its operational procedures and move back to initial implementation stage. This book accommodates the nonlinear transition of program stages and the evaluations related to each stage. Readers can pick and choose from the chapters and rearrange the sequence of program stages to fit their programs.

Classes of Strategies and Approaches
Serving Stakeholders' Various Needs

An evaluation strategy is a general direction or orientation the evaluator and stakeholders take in order to fulfill a given evaluation's purpose. For example, merit assessment is one general strategy employed to serve stakeholders' accountability needs. Other evaluation strategies include the *development* and *enlightened* strategies. The overall evaluation strategy must be closely related to the stakeholders' evaluation needs. In contrast, the *evaluation approach* constitutes a systematic set of concrete procedures and principles to guide the designing and conducting of an evaluation. The evaluation approach determines the evaluation's focus; it affects the research methods applied to collect and analyze data, as well as the interpretation of data. Typically, several evaluation approaches are harnessed together within one evaluation strategy (see Table 3.1). Well-known evaluation approaches include experimental and quasi-experimental approaches, monitoring, needs assessment, and qualitative evaluation.

Whereas most stakeholders are unfamiliar with individual research procedures and techniques (familiarity with which is assumed of evaluators), they are usually acquainted with the general directions that evaluation strategies comprise. For example, stakeholders may not know what a quasi-experiment is, but they do understand generally what a strategy such as merit assessment entails. The easiest and best course for evaluators is to determine the appropriate program evaluation approach by communicating with stakeholders about the strategies the evaluators think will fit the stakeholders' evaluation needs. With their input in mind, the evaluators can then lead a discussion of various appropriate evaluation approaches. Too often, evaluators ignore the dialogue on evaluation strategies, launching right into selection of evaluation approaches. The fallout from this practice can be stakeholders' *un*informed consent to employ whatever evaluation approach the evaluator recommends. With little or no understanding of what that approach consists of, stakeholders may, when handed the final report of the evaluation, realize that it is not what they wanted and does not provide the information they need.

My view on evaluation strategies and approaches has been greatly influenced by my acquaintance with the following case of misdirected evaluation. The client was a group of high-performing community-based organizations seeking to provide capacity-building services to similar but less accomplished organizations. A skilled and respected evaluator carried out the project. Before beginning, this evaluator met with stakeholders several times to discuss potential evaluation approaches. The parties decided to adopt mixed methods to assess the merit of the capacity-building program. The final evaluation report provided a detailed pros-and-cons assessment of the program, expressing in general that the evaluation favored the project. Unfortunately, those anticipated to be the program's service providers complained that the evaluation offered few insights into improving their program. The generated information was of the wrong kind, they protested, because the evaluation failed to reflect their needs and views. In the end, it became clear that the service providers had wanted a development-oriented evaluation, whereas the evaluator had conducted a merit-assessment evaluation.

This is not a case easily dismissed by blaming the service providers for misstating their evaluation needs at the beginning, or for changing their minds later on. In reviewing the project with the evaluator and the service providers, it appears that the heart of this problem was the absence of effective tools with which to voice actual evaluation needs and identify suitable accompanying evaluation approaches.

A superabundance of evaluation strategies and approaches is cited in the literature (e.g., Mark et al., 2000; Rossi et al., 2004; Shadish et al., 1991; Stufflebeam, 2001). This practical taxonomy (Table 3.1) does not include every strategy and approach ever conceived. Evaluation strategies and approaches discussed frequently in the evaluation literature are not necessarily those used frequently by evaluation practitioners, and vice versa. In contrast, the practical taxonomy of program evaluation means and ends includes only strategies and approaches with marked potential to be used successfully in the field regardless of its popularity in the existing literature. Readers for whom the terminology in Table 3.1 is new should note that subsequent chapters of the book explore and explain the strategies and approaches in detail. The bird's-eye view provided in this chapter is simply a foundation.

To begin, there are four general categories of evaluation strategy included in the practical taxonomy of program evaluation means and ends: *merit assessment, development, enlightenment,* and *partnership.*

Merit Assessment Strategies

Merit assessment strategies are those that can provide information on the performance or merit of a program. Two merit assessment strategies frequently used by evaluators are the *performance assessment strategy* and the *performance monitoring strategy.* (Only a brief introduction to these follows here; Chapters 8 and 9, however, discuss at length the differences and relationships between performance evaluation and performance monitoring.) *Performance assessment* is the employment of rigorous designs to provide credible information about a program's merit in terms of either its implementation process or its outcomes. The performance assessment strategy is part of a long-standing, influential tradition in program evaluation, one discussed in Mark et al. (2000) and Shadish et al. (1991). It is often more expensive and time-consuming to employ the performance assessment strategy than other comparable evaluation strategies. The most common application of performance assessment strategy is with programs in the mature implementation stage or outcome stage. Performance assessment strategy is commonly affiliated with the use of a popular evaluation approach called *fidelity evaluation*. The fidelity evaluation approach assesses whether a program has been or is being implemented according to expectations. Another approach popularly used with the performance assessment strategy is the *outcome evaluation*, which assesses a program's success in reaching its goals. The outcome evaluation approach is the major tool within traditional outcome evaluation (e.g., Cook & Campbell, 1979).

The *performance monitoring strategy*, the second assessment strategy included in the practical taxonomy, involves using indicators to follow the implementation process and outcomes of a program across time. In a drug treatment program, for example, evaluators might monitor clients' drug use both before and after they experience the intervention. The performance monitoring strategy comprises two well-known approaches: *process monitoring* and *outcome monitoring*. Process monitoring cannot produce in-depth information about a program's implementation as process evaluation does; neither is outcome monitoring likely to produce convincing data about an intervention's effect on outcomes the way outcome evaluation can. In their defense, however, process monitoring and outcome monitoring are useful for managing a program and likely to cost less than typical process evaluation and outcome evaluation.

Development Strategies

Development strategies collect evaluative data relatively quickly in order to assist stakeholders with program planning or development. Three development strategies are well established in program evaluation: the *background information provision strategy*, the *troubleshooting strategy*, and the *development facilitation strategy*.

Background Information Provision Strategy. To use the background information provision strategy is to research the background of a program in terms of community characteristics and needs, target population characteristics, and/or intervention options. The information gathered should help program designers and other stakeholders plan or strengthen a program. Evaluation approaches suited to this strategy include needs assessment and formative research. Determining and prioritizing the needs of a community or target population that warrant intervention is called needs assessment, such as when an agency asks what kinds of youth services are most needed in a community. In such a case, program evaluators might systematically interview youths, parents, and community leaders to help the agency answer its question. Formative research differs from needs assessment in the greater emphasis it places on an identification or prioritization of needs. Formative research consists of gathering empirical information on community and target population characteristics, as well as intervention options, in order to help stakeholders plan and develop programs. For example, program designers uncertain about what kind of drug prevention program would be best received by new immigrants might engage program evaluators to manage a survey or focus group meeting, obtaining information that enables the program designers to make a decision.

Troubleshooting Strategy. The *troubleshooting strategy* is a system for identifying trouble spots in programs and addressing them. The troubleshooting strategy is used, first of all, to provide timely assessment of barriers and/or problems facing a program; its second use concerns options available to stakeholders to address difficulties. The value of the strategy lies in its ability (not always guaranteed) to effectively identify any implementation problem *before* it gets away from stakeholders and major damage occurs. Evaluators using this strategy must also provide stakeholders with information that facilitates resolution of the problem. The troubleshooting strategy is associated with use of the *formative evaluation, relevancy testing, pilot-testing,* and *commentary and advisory* approaches. *Formative evaluation* is associated with research methods that are flexible to use; are easy to adopt in the field; and have a short turnaround time, such as focus groups and participant observation to collect, in timely fashion, facts about barriers and problems in implementation that promise to strengthen it. Having chosen to target newly arrived immigrants, for instance, an HIV prevention program further decides to serve them with group counseling. After the implementation is carried out, evaluators are contracted to look for potential problems in the recently completed process. Using formative evaluation, evaluators interview a sample of the clients and quickly learn that some clients—Asian immigrants—are uncomfortable in group discussions of sexual behavior. The quick feedback made available to the program director by the formative evaluation approach prompted modification of the program to better serve this particular immigrant group.

Formative evaluation and formative research (an approach affiliated with the background information provision strategy) are both research activities, yet with an important difference: Whereas formative evaluation examines directly the program's implementation, formative research is usually carried out before implementation and produces background information related to program planning. For example, the evaluator tackling the above assignment from a formative research approach might study a target group's cultural background as it relates to sexual behavior in hopes of facilitating program design decisions. An evaluator using the formative evaluation approach would evaluate the given target population's experience with the program itself.

The troubleshooting strategy also includes *relevancy testing.* Relevancy testing is the small-scale assessment of causal assumptions underlying a program and whether these assumptions hold up in the field. Relevancy testing can be used to strengthen the soundness of a change model. In contrast, the *pilot-testing approach* to the troubleshooting strategy involves actually operating the program on a very small scale. Unlike relevancy testing, pilot testing usually focuses on the action model. The information and experience gained from pilot testing can help strengthen a program before formal implementation begins because areas needing modification can be fixed early and prevented from affecting the full-scale implementation. Another troubleshooting strategy is the *commentary and advisory approach,* which is an approach that does not collect data from the field. Instead, the expertise of evaluators is tapped as they review and comment on an existing action model and change model. They advise stakeholders about probable strengths and weaknesses of the models and offer suggestions for improvement. Finally, the *program review/development meeting approach* generates insights through systematic discussions in a meeting format among a group of program implementers and staff. With the evaluator providing facilitation, the experiences of a program are discussed, any implementation problems are dissected, barriers to and facilitators of these problems are identified, and strategies are created to strengthen the program.

Development Facilitation Strategy. Evaluators' knowledge and skills are also central to the *development facilitation strategy,* which is defined in this book as the use of such expertise to help key stakeholders in a meeting/workshop setting. The development facilitation strategy functions to facilitate the stakeholders' efforts to develop or fine-tune the logic of a program, or to identify its problems and seek programmatic solutions for them. Using this strategy, evaluators become facilitators and consultants, essentially; Guba and Lincoln (1989) and Patton (1997) have emphasized the value of this method for solidifying a common vision, winning support, and broadening a program's capacity.

Expert evaluators can draw on their program evaluation skills to contribute greatly to the development of coherent programs that are logical in their foundations and feasible to implement.

Some evaluation approaches associated with the development facilitation strategy are the *conceptualization facilitation approach* and the *concept-mapping approach*. The conceptualization facilitation approach requires evaluators to work as facilitators and consultants, clarifying stakeholders' ideas, especially those concerning action and change models—and then, quite likely, facilitating their effort to develop these models. Under the concept-mapping approach, quantitative methods provide help to stakeholders sorting out the ends and means of a program (Trochim & Cook, 1992).

Enlightenment Strategy

Stakeholders may, of course, seek program evaluation in response to accountability, as well as program improvement, needs. As discussed in Chapter 1, program improvement remains the ultimate goal of program evaluation, and pure performance evaluation has little to say about improving programs; yet it is possible to design evaluations to meet both kinds of needs. The key is extending the evaluation beyond merit assessment by examining underlying assumptions and mechanisms that mediate the effects of the program. Evaluators with this orientation are practicing the *enlightenment strategy*. Enlightenment strategy is discussed at length in the literature (e.g., Mark et al., 2000), and, in general, it takes the position "Assessment is means, program improvement is end." The work by Cronbach (1982), Chen (1990), Chen and Rossi (1992), and Shadish, Cook, and Levinton (1991) is related to enlightenment strategy. An evaluation approach that suits this strategy is called the *theory-driven* (or *theory-based*) *evaluation approach*. Theory-driven evaluation has been applied to assess implementation processes as well as program outcomes (Chen, 1990).

Partnership Strategy

The final strategy presented in this section of the book is the *partnership strategy,* in which stakeholders invite evaluators to be partners in planning and implementing programs. The parties work together closely at every step, with evaluation information introduced regularly to support their effort to develop and implement a program. This strategy, and the *bilateral empowerment approach* that accompanies it, comprise something of a challenge to the traditional foci of evaluation. Bilateral empowerment means that the participating evaluators are granted membership on the development team. Accordingly, they have direct input as to the handling of development and evaluation issues; that is, evaluators participate in the decision-making process. Bilateral empowerment may work best with programs that have vague notions about goals, interventions, and implementation. This strategy and this approach have gained momentum in the recent literature concerning community coalition evaluation (e.g., Goodman, Wandersman, Chinman, Imm, & Morrissey, 1996).

STEPS TO TAKE IN APPLYING
THE PRACTICAL TAXONOMY

The purpose of the practical taxonomy of evaluation ends and means is to associate particular evaluation strategies and approaches with particular program stages and stakeholder needs (see Table 3.1). The stakeholders of a program in its initial implementation stage, for example, need an evaluation strategy and approach that move quickly to tackle immediate implementation problems. The taxonomy demonstrates a very clear bearing: that *program evaluation is situational.* No single evaluation strategy, approach, or method can succeed with every possible evaluation need or situation. Means of evaluation that are fruitful in one case may be fruitless—or even misleading—in others. The performance assessment strategy, for instance, although plainly useful when the need is for accountability of a program in its mature implementation stage, could produce questionable results if employed with an immature program. This is because the only input stakeholders can actually use early on is timely information that helps to stabilize early implementation.

The practical taxonomy as it appears in Table 3.1 was crafted as a "map" of the art of evaluation for evaluators and stakeholders to review together. An evaluator might want to proceed through the taxonomy (from left to right) with stakeholders, identifying the evaluation strategies and approaches best suited to the evaluation they seek. Taking the following steps *in sequence* should bring the evaluator to the finish line in good shape.

1. *Identify the program stage that is of interest.* Evaluation needs are expressed by stakeholders in general, abstract terms. The evaluator must create precision in the discussion by facilitating a choice about exactly which program stage(s) should be the focus of investigation. When stakeholders request evaluation of a program implementation, they must decide if they mean its initial implementation or its mature implementation, because the two are not identical. Lack of expressed *stage-specific* needs, understood both by stakeholders and evaluators, can end in the choosing of mismatched strategies and approaches, producing a useless evaluation. Stakeholders cannot be blamed for misunderstandings about evaluation needs because it is the evaluator's responsibility to thoroughly grasp stakeholders' intentions *before* designing an evaluation. Information obtained in the course of articulating or clarifying stakeholders' needs will advance the effort to select the best evaluation strategy and approach for the task.

2. *Choose an evaluation strategy that matches stakeholders' internal/ external purposes.* Having settled the issue of program stage, the evaluator must quiz stakeholders about the eventual audience for the evaluative information. Does it have an internal purpose, external purpose, or both? This is crucial when selecting an evaluation strategy. In general, if the information mainly will be used *internally* to find and fix implementation problems, then the development facilitation strategy is a good choice. For example, stakeholders desiring to troubleshoot their programs will find formative evaluation to be valuable. For an audience beyond the internal, however, evaluators and stakeholders might use an assessment strategy because assessment strategies provide much information that satisfies accountability requirements. A performance assessment strategy used at the outcome stage, for example, can be used to rigorously assess the effects of a program. But should the stakeholders need evaluative information that serves program improvement needs as well as accountability needs, then the enlightenment strategy is the best choice.

3. *Choose evaluation approaches and research methods that provide acceptable trade-offs.* With the strategy question answered, it is time to choose an appropriate, stage-specific evaluation approach (or approaches). Each strategy included in the practical taxonomy is linked to one or more evaluation approaches, and each of those is, in turn, affiliated with a number of research methods. All of these options demonstrate strengths and weaknesses in terms of the basic qualities of evaluations: timeliness, rigor, thoroughness, and cost.

Stakeholders must be willing to make trade-offs among these qualities, with adequate understanding of pros and cons. Two examples of trade-offs are illustrative. First is the forming of an acceptable compromise concerning the timeliness, rigor, and cost of evaluation. We accept that there is a tendency among evaluators (or at least a desire) to take whatever evaluation approach is the most rigorous. Rigorously designed evaluations with stringent methodologies are likely to be accepted by the scientific community and perhaps published in prestigious journals. But rigorously designed evaluations with stringent methodologies are usually expensive, and stakeholders may not have the necessary funds. Similarly, rigorous designs are not completed quickly, and stakeholders may be working within a window that accommodates client or community needs rather than scholarly ones. To make a generalization, the evaluation approaches and research methods within the assessment and enlightenment strategies of the taxonomy demand more scientific rigor and so take more time to finish. On the other hand, the evaluation approaches under the taxonomy's development strategies, although they manifest a brevity that loves deadlines, also embrace "flexible" methods like the focus group, which can be construed as departing from the rigor of the established scientific standard. This book certainly endorses the use of rigorous designs and methods where and when feasible. It equally reiterates that program evaluation is an applied science. Serving stakeholders' needs as responsively as possible must remain a paramount concern as the evaluation approach and research method are selected. Rigor is *a* major factor, not *the* major factor, for the evaluator's consideration. So, if stakeholders offer sufficient money to support rigorous designs and methodologies, evaluators should exploit this. When money or time is necessarily limited, however, evaluators should not feel compelled to advocate an evaluation approach and research method that would be a financial burden or come to its conclusions belatedly.

Here is an example. The methods of outcome evaluation (such as efficacy evaluation or effectiveness evaluation) are rigorous and lengthy, whereas those of outcome monitoring are less demanding. Stakeholders whose priority is highly credible and precise information about a program's effects want outcome evaluation. Stakeholders on a tight budget of cash, time, or both, want something else. If they want simply some rapid feedback about clients' progress, it would be inappropriate for an evaluator to advocate an expensive outcome evaluation when less costly outcome monitoring can also provide that feedback. The evaluator's role is to inform stakeholders that such an option exists, and that it represents a trade-off, but one that will conserve their time and money. A second illustrative example lies in the trade-off between cost and thoroughness. Evaluative information can be costly; the deeper an evaluation delves, the costlier it becomes. Programs almost always are constrained by cost, and the results, when they are evaluated, are a trade-off between evaluative product and price. Stakeholders with a program in the planning stage need to realize that they can save money by seeking only an evaluator's comments on a program plan *if* they can forgo the deeper data that costlier formative research or needs assessment would provide. Of course, if evaluator comments are unlikely to shed any new light on the program plan, the stakeholders might be better off waiting until they can afford the more expensive option and its data bearing directly on the decisions they must make.

4. *Communicate to stakeholders facts about the chosen evaluation strategy/approach and research method.* When the evaluator has determined which evaluation strategy, evaluation approach, and research method fit the assignment best, he or she must explain them carefully to the stakeholders. Stakeholders should be especially well instructed about the kind of information that will be the final product. Communication helps prevent misunderstanding between stakeholders and evaluator. It gives stakeholders an opportunity to voice any doubt about the proposed evaluation's capacity to meet their needs. (Any evidence of such doubt should cause the evaluator to reexamine the options.) Finally, free communication with stakeholders also gives evaluators a forum for detailing the kind of support expected from stakeholders throughout the evaluation process.

EVALUATION RANGING
ACROSS SEVERAL PROGRAM STAGES

Program evaluators are frequently engaged to conduct multiple-entry evaluations; that is, evaluations across program stages. Before beginning, conflicts of interest that could be created by such multitasking must be addressed. Generally speaking, when the various tasks all fall within the domain of the development strategies, or the domain of the assessment/enlightenment strategies, conflict of interest is negligible. Evaluators can, for example, carry out evaluation activities that assist in the development of a program plan and also, later on, provide the data to facilitate program implementation. Because each evaluation is confined to one phase and thus is of a consistent nature, the evaluations complement each other instead of competing with each other. Similarly, no conflict results when an evaluator performs assessment evaluation during the implementation stage and goes on to assess the program's effectiveness in the outcome stage. The natures of the two evaluations are compatible.

Attention to conflict of interest is warranted when evaluators doing development-oriented work with programs in their early stages subsequently become responsible for assessing program performance/merit in later stages. Whether an actual conflict exists depends on the strategies and approaches involved and on whether evaluators had a direct role in the decisions made about program planning and implementation. Conflicts of interest are quite likely to occur when evaluators conduct bilateral empowerment evaluation, becoming active members of design/development teams (as in the development partnership strategy described above), then later assume responsibility for assessing program merit. A team member-evaluator is seen as having a vested interest in the program. If he or she were to declare the program successful, the credibility of the outcome could well be suspect. Following completion of empowerment-based evaluation projects, it is much better to secure new evaluation professionals to carry out any assessment or enlightenment type of evaluation.

Evaluators whose involvement in the development facilitation strategy is limited to facilitating the work of stakeholders are not prohibited from conducting assessment or enlightenment types of evaluation of the program during later stages. "Vested interest" is not applicable in cases in which evaluators conducted needs assessment, formative research, or formative evaluation (in the program planning stage) for the benefit of *stakeholders* designing or developing their program. This facilitation experience is not grounds to exclude them from evaluating the program's implementation and effectiveness later on. In the same way, an evaluator who has worked to facilitate stakeholders' development of a logic model or program theory is not barred from later assessments of the program. However, as a precaution protecting the perceived credibility of an evaluation, evaluators in these situations need to do three things. They must first offer up for discussion and scrutiny the fact and the nature of their earlier involvement in development activity. Second, they must make it clear to stakeholders that the requirements for evaluating programs in later stages differ from requirements for development-oriented evaluations. As a final condition, they must document explicitly how they arrived at the major conclusions of their evaluations.

HELPING STAKEHOLDERS GEAR UP (OR CLEAR UP) THEIR PROGRAM THEORY

As the practical taxonomy suggests, when evaluators set about reviewing a program using an approach associated with development facilitation strategy or enlightenment strategy (strategies discussed further in the chapters that follow), a frequent first requirement is clarification of the stakeholders' program theory (Chen, 2003). At times, the evaluator may even need to help the stakeholders with the initial draft of a program theory. This section explores ways to clarify or help develop stakeholders' program theories.

Reviewing Existing Documents and Materials

To start the process, evaluators need to study existing documents or materials related to the program—brochures, pamphlets, grant applications, memos, and so on. This provides general information, preparing the evaluator for subsequent interviews with stakeholders and ensuring that these will be conducted efficiently. Evaluators might also consider visiting program sites to increase familiarity with programs that have already been implemented.

Clarifying Stakeholders' Theory

As the evaluator begins to clarify stakeholders' program theory, or as such a theory begins to be developed by stakeholders with assistance from the evaluator, an important issue must be resolved. What role should the evaluator play in this process? How can he or she best contribute to the work? The evaluator should remember that a program theory *belongs to the stakeholders;* the evaluator's function is that of facilitator and consultant. Evaluation skills and knowledge should be brought to bear to increase the productivity of the meetings at which various stakeholders attempt to articulate and refine their ideas about the program theory. Stakeholders are sure to have divergent backgrounds, concerns, and interests. It is easy for them to eat up time with freeform discussions that never even approach agreement. The evaluator's job as facilitator is to outline for the group the salient issues to discuss, showing stakeholders where to fill in with their own experiences, thought, and expertise. Next, the evaluator can synthesize the discussions and build consensus. The evaluator's concurrent job as consultant means filling in with his or her own evaluation expertise when stakeholders ask for advice. The evaluator is present to lay out options for stakeholders to consider and should avoid imposing his or her own values upon stakeholders. The evaluator should also present ideas drawn from his or her own expertise for stakeholders to discuss.

Participatory Modes for Development Facilitation

Evaluators can assist stakeholders whose program theory is under development by adopting either of two general participatory modes: the *intensive interview mode* or the *working group mode*. Choosing a mode is a prerequisite for stakeholders and evaluators preparing to work together. The *intensive interview mode* centers on individual, intensive interviews that the evaluator holds with representatives from each key stakeholder group. The aim is to record systematically the individuals' perceptions about issues within the incipient program theory. Based upon these interviews, the evaluator formulates a first draft of the program theory, which will be read by the representatives and other stakeholders. Their comments are considered as the final draft is prepared. In addition, evaluators can meet with these individuals for the purpose of fine-tuning and finalizing the program theory. The *working group mode* similarly involves representatives from key stakeholder groups. However, in this mode, the representatives are not interviewed individually but instead meet together with the evaluator to develop the program theory. Group members need to consist of those who will be most deeply involved in formulating and designing the program, those who will be most deeply involved in implementing the program, and other key constituencies whose input will be influential as to the direction the program will take. The facilitator, of course, is another member.

This list actually creates relatively few participants when the planned program is a small one. With large programs, however, the working group tends to be too large. A group that is too large can discourage members' full participation, at the same time necessitating many more sessions to finish the work. A good rule of thumb is to limit a group to no more than 15 members. Small groups can foster a casual atmosphere for discussion, enabling the evaluator to serve as both facilitator and consultant. A large group, especially one with a highly diverse and vocal membership, makes it difficult for the evaluator to be facilitator and consultant at once. With large groups, at least two evaluators may need to participate in the meetings—one as facilitator, the other as consultant.

How should one choose a participatory mode? Each has its advantages. The intensive interview mode tends to be less challenging logistically because group meeting arrangements are needed only infrequently. In addition, the interview setting may strike some participants as being much more comfortable and secure than a typical meeting. The interview also tends to better promote probing of stakeholders' views by the evaluator. A potential limitation of the intensive interview mode, however, is some stakeholders' perception that they have participated in only one part of the theorizing process. This is especially problematic in large programs with many powerful stakeholders. In contrast, the *working group mode* tends to demonstrate that the program theory is being developed in an open, inclusive manner, which could increase some stakeholders' buy-in. But again, work with a group often requires more time to finish than work done in interviews. Furthermore, it is possible in working groups for a few highly vocal stakeholders to dominate discussion. This problem might be alleviated if the evaluator sets clear rules of discussion from the first meeting. Rules should encourage full participation by all members. An even more serious problem with the working group mode is that some stakeholders—those in the lower ranks of the implementing organization(s)— may worry about expressing their actual opinions, choosing instead to simply echo what higher-ranking officials say. In such a case, the final program theory could reflect only the views of those in authority. If this is a concern, the intensive interview mode is the better choice.

Theorizing Procedures for Development Facilitation

As with the participatory mode, a *theorizing procedure* must be selected in order to help stakeholders develop their program theory. So-called *forward reasoning, backward reasoning,* and *forward/backward reasoning* are the three general options for evaluators working within the development strategies. Backward reasoning is an approach that begins with the change model, then moves backward step by step to the action model in order to obtain the program theory. It is "backward" reasoning in that the process moves in the direction opposite of the sequences shown in Figure 2.1. More specifically, backward reasoning starts from the question of what goals the program seeks to achieve. Other questions are the following: On which determinants of these goals should the program focus? What intervention will affect these determinants in appropriate ways? When a change model has been completed, evaluators can facilitate stakeholders' development of the corresponding action model with questions such as these: Which groups need to be reached and served? What kind of program implementers and implementing organizations will suit? What types of intervention and implementation protocols seem best? Should there be collaboration with other organizations? Will the program require ecological support?

Forward reasoning, on the other hand, means formulating a program theory in accord with the logic flow outlined in Figure 2.1—action model first, then change model. Forward reasoning produces general program goals and grows from initial thought on what kind of action model is needed. Questions like these are important in forward reasoning: At which intervention and implementation protocols will the implementing organizations excel as they try to solve particular problems or reach certain goals? What group needs to be reached with the intervention, and *how* can it be reached? What setting and delivery mode make sense? Do clients face barriers to receiving services, and can the program alleviate these? How and where should contextual support be sought for the intervention, if needed? When they have completed the action model, evaluators and stakeholders can develop a change model by asking two questions, in sequence: What determinants will be changed by the intervention? What outcomes will be achieved by changing these determinants?

Forward reasoning and backward reasoning alike can be used successfully in the formulation of program theories. In certain circumstances, however, one of the two theorizing procedures is clearly the better choice. Some rules of thumb can guide the evaluator.

The first rule says that, generally speaking, when program designers and other key stakeholders are familiar with social science methodology, backward reasoning works best. It is the procedure that starts with discussion of a program's goals, a subject stakeholders enjoy discussing and that can help break the ice. Subsequent inquiries within the backward reasoning procedure (e.g., What are the causes of the problem? Which intervention seems to offer promise? What is an appropriate design for the intervention?) are well within the stakeholders' capability to debate. On the other hand, when program designers and other key stakeholders are not familiar with social science methodology, forward reasoning should be preferred. The reason is that theorizing procedures need to start with a topic that stakeholders feel comfortable discussing. Forward reasoning starts with the specification of programming issues, about which stakeholders have many ideas to voice. Forward reasoning aptly suits efforts to clarify or develop stakeholders' views on the actual steps their program should take: what to do first of all, what to bring in next, building up to the third and fourth and fifth steps, and so on through culmination in delivery of a service or services. Whether an evaluation begins with forward or backward reasoning, if the evaluator and stakeholders come to realize that continuing in that mode will be difficult, they are always free to switch to the other procedure to resume their discussions.

It is also important to note that forward and backward reasoning are not mutually exclusive. The forward/backward reasoning is a use of forward and backward reasoning, back and forth, to facilitate stakeholders and make explicit their program theories. The forward/backward reasoning is more time-consuming than the other two approaches, but may have the best of both worlds. In using this technique, evaluators and stakeholders often apply backward reasoning first and then use forward reasoning to compensate for weaknesses in backward reasoning. For example, an evaluation focused on *both* action and change models might begin with the forward reasoning procedure to construct an action model, take up backward reasoning to establish a change model, and finally integrate the two to arrive at an overall program theory. This dual procedure is a good choice when program stakeholders and evaluators believe that unintended outcomes will be of import. Employing the theorizing procedures in both directions may make it more likely that a working group will be alerted to potential unintended desirable or undesirable effects. The evaluator should facilitate discussion of any unintended effects and their prevention, should they be undesired.

Preparing a Rough Draft that Facilitates Discussion

The act of developing a useful program theory is often time-limited. The work's usefulness may dwindle with the passing of a deadline and, more often than not, deadlines come sooner than is desirable for the planning team. Scheduling, preparing for, and executing either interviews or meetings, and then compiling the information obtained and soliciting comment on it, is very time-consuming (and especially so if every element and issue needs to be broached, examined, and ruled on—from scratch—in these meetings or interviews). To shorten the period required, it is not unusual for evaluators to scour existing information about a program and use what they learn to prepare a rough draft of a program theory for discussion by the working group. The rough draft should include the elements of a program theory stated in the existing information, the elements that may be implicit in the existing information but are not communicated straightforwardly there, and the significant elements not yet touched on that will require intensive discussion. The rough draft provides a focus for stakeholders' thoughts and suggestions. It should be distributed to members of the working group (or to individuals scheduled for interview) well in advance of the meeting date, giving them time to digest the contents. The rough draft is a tool to streamline discussion, focus comment, and foster specificity and usefulness in the work.

DETAILS OF THE WORK:
DESIGNING AN EVALUATION

The taxonomy displayed in Table 3.1 and discussed above is useful for identifying stakeholders' evaluation needs at certain program stages. That task concluded, evaluators may consult Chapters 4 through 10 for the detailed principles and guidance helpful in making a final choice of evaluation strategy and approach. With a strategy and approach (or approaches) determined, the time has arrived to design the evaluation and launch it in the field. Chapters 4 and 5 (and beyond) discuss evaluation design in depth and are tailored to cases in which stakeholder needs pertain to the planning stage. Chapter 6 concentrates on designing evaluations at the initial implementation stage, and in Chapter 7 (and in the process-monitoring portion of Chapter 8) the focus is evaluation design at the mature implementation stage. The outcome-monitoring portion of Chapter 8, and Chapters 9 and 10 as well, are tailored for evaluators involved in outcome stage evaluation.

DYNAMICS OF EVALUATION ENTRIES
INTO PROGRAM STAGES AND HOW TO APPLY
THIS BOOK TO CONDUCTING EVALUATION

The application of evaluation along program stages is dynamic in nature. Evaluators might be asked to conduct an evaluation focusing on either any one

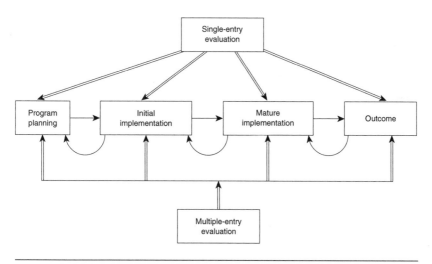

Figure 3.1   Single-Entry Evaluation Versus Multiple-Entry Evaluation

stage or a combination of stages. Figure 3.1 highlights the dynamics of such evaluation application.

The center section of Figure 3.1 indicates that program stages ideally move from planning to initial implementation to mature implementation and, eventually, to outcomes. It is possible, however, to move in a nonlinear fashion. Figure 3.1 demonstrates both single entry and multiple entry evaluation. The definitions of these two types of evaluation, as well as how to apply this book to conducting these two types of evaluations, are presented below.

1. Single Entry Evaluation

In single entry evaluation, evaluators focus their evaluation on a single program stage. This book is organized in a way that accommodates a single entry evaluation. As long as readers have a basic knowledge of the information found in Chapters 1 through 3, they can go directly to the chapter that applies to the stage of evaluation in which they are interested. For example, if evaluators are interested in outcome evaluation, they can move from Chapter 3 directly to Chapters 9 and 10, which discuss the major issues of outcome evaluation; they do not have to refer to Chapters 4 through 8 in order to conduct an outcome evaluation, though these chapters may still prove helpful. Similarly, readers that are interested in program planning can move from

Chapter 3 to Chapters 4 and 5, which discuss evaluation approaches used in the planning stage. Readers interested in conducting an evaluation at the initial implementation stage can move from Chapter 3 to Chapter 6; readers interested in process evaluation can move from Chapter 3 to Chapter 7; readers interested in program monitoring can move from Chapter 3 to Chapter 8; and readers interested in outcome evaluation can move from Chapter 3 to Chapters 9 and 10.

2. Multiple Entry Evaluation

In multiple entry evaluation, evaluators are concerned with conducting an evaluation that focuses on two or more program stages. For example, at the beginning of a program, evaluators may be asked to conduct an evaluation of any two or more program stages, from planning to outcome. Similarly, in an established program, evaluators may be asked to conduct an evaluation covering both the implementation and outcome stages. This book can be used effectively to guide multiple entry evaluations. After evaluators and stakeholders have decided which combination of program stages or evaluation approaches are to be used, evaluators could read the chapters relevant to these stages. For example, if evaluators are asked to conduct process evaluation and outcome evaluation, they could refer to Chapters 7, 9, and 10. Similarly, if they are asked to conduct evaluation in program planning and initial implementation stages, they could refer to Chapters 4, 5, and 6.

The entries of program stages in a multiple entry evaluation could be nonlinear. For example, imagine that stakeholders are not happy with their existing program. They may ask evaluators to conduct an evaluation at the mature implementation stage to learn from their mistakes and then ask evaluators to conduct evaluations at the planning stage to facilitate their development of a new program. In this case, the evaluators could refer to Chapter 7 first and then to Chapters 4 and 5.

PART II

PROGRAM EVALUATION TO HELP STAKEHOLDERS PLAN INTERVENTION PROGRAMS

———•◦•———

The scope of program evaluation is now much expanded from traditional areas such as outcome and implementation. With growing awareness that implementation of programs is affected by the quality of program planning, a further role was revealed for evaluation. The strategies and approaches of program evaluation foster the kind of understanding most likely to ensure top-quality program planning.

Generally speaking, the earlier that program evaluation techniques are incorporated in the planning of a program, the easier it becomes for the directors and implementers to improve the new program using evaluation feedback. Program staff can modify a program much more readily during the planning stage. Once a program is established and on its way to becoming routine, enacting substantive changes can be difficult, even when evaluation results strongly support them. Program evaluators who are serious about putting their evaluation results to work need to learn how to apply evaluation strategies and approaches to assist stakeholders in program planning and development. The first requirement of a sound program plan is the well-developed program rationale, which is the topic of Chapter 4. Chapter 5 takes up the preparation of the program plan itself.

FOUR

ASSISTING STAKEHOLDERS AS THEY FORMULATE PROGRAM RATIONALES

———•◦•———

The design of an intervention program is the responsibility of program designers and other key stakeholders. Knowing how important—and complicated—the planning of a program is, these parties will frequently seek expert help with the planning process. They are most likely to benefit from consulting an evaluator in an effort to conceptualize the program more soundly. This happens, of course, at the program planning stage (as illustrated by the figure of the taxonomy; see Table 3.1). This chapter was written with start-up programs uppermost in mind, but its content has implications for established programs as well. Because it is not unusual for established programs to experience changes in policy, clients, personnel, management, and/or leadership over the years, stakeholders may periodically feel that they have lost sight of their program and need to reconceptualize it, with help from the evaluator.

THE CONCEPT OF A PROGRAM RATIONALE AND ITS PURPOSES

As they plan a program, stakeholders face two crucial tasks: articulation of a program rationale and development of a program plan. The rationale and plan together provide the program's infrastructure. The program rationale issues a

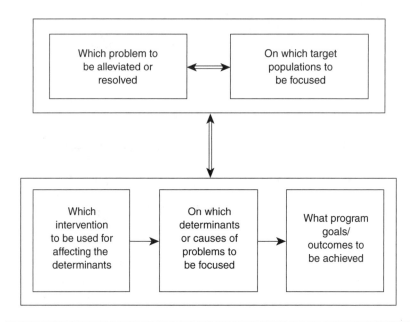

Figure 4.1   Conceptual Framework of Program Rationale

call for action, and the program plan is a blueprint of that action. In this chapter, the program rationale is central, whereas the program plan is featured in the next chapter.

The form the program rationale takes is that of systematic argument in which stakeholders assert the potential of a specified intervention's potential to achieve a stated goal. In the program rationale, the goals to be pursued are named and reasons are given for the selection of these goals. In addition, the rationale names the intervention to be used and explains how it should lead to the goal. Finally, the program rationale specifies the target group to be serviced by the intervention and explains why the group requires it. Until recently, program designers and stakeholders typically focused more intensely on producing a program plan than a program rationale. A change has been in the air, however, because increasing numbers of stakeholders have started asking evaluators to become involved in programs' initial stages. The art of program evaluation is proving very useful here, too. Figure 4.1 illustrates a conceptual framework that typically underlies articulation of a program rationale.

From the framework's component concepts are born questions like these: What problem will the program work to alleviate or resolve? What population will receive the program's attention? At what measurable goals will the program aim? On which determinants creating change should the program focus? What interventions can affect the determinants?

Figure 4.1 shows the component concepts grouped in two boxes. The top box displays two components—the identified problem and the population targeted for services. The bottom box contains each component of the change model. Two-way arrows link the target group and the change model to demonstrate the importance of the "fit" between change model and target group. In other words, it is vital that goals, determinants, and interventions (the three change model components) are appropriate to the target population and the problem it faces. Figure 4.1 also serves to model the kind of diagram that is almost always helpful when presenting a finished program rationale to its audience. It may also be acceptable, at times, to present the rationale simply as a written statement.

The interwoven nature of the program rationale and program plan means that the two are usually developed together. Nevertheless, they do differ conceptually. The program rationale focuses on the change model, whereas the program plan is concerned with the action model. There are, in fact, advantages to be gained from developing the program rationale fully before attempting the program plan.

Why a Program Rationale?

Programs rely on their program rationales for a *foundation* for planning, for efficient *communication,* and for a basis of outcome *evaluation.* Program plans must be developed systematically. One major weakness in contemporary program planning is that program designers tend to create plans in a rush. With just a passing glance at the program rationale, too often they try to proceed, even though they lack a clear vision. In consequence, their plans tend to be unfocused and disorganized, and they are unlikely to be effective.

A well-articulated program rationale can give program designers a firm foundation for their efforts. It is an outline they follow, a constant reminder of what makes program activities meaningful and prone to achieving goals. Furthermore, the program rationale can give insight into the way individual efforts support the overall program mission. A good program rationale also fosters efficient communication between program director, program staff, and audiences within and outside of the program. It is common today for program designers to use lengthy narrative descriptions to introduce people to their programs. These descriptions may take too much time to read, bogging down in details and straying from things people really need to know, like what the program seeks to do. In contrast, the program designer with a well-articulated program rationale has at hand a concise, comprehensible, introductory summary of the program, one useful for facilitating communication. Yet another function of a sound program rationale is the establishment of a basis for later evaluation of program outcomes (see Chapters 9 and 10 for an intensive discussion of this function). Containing as it does statements of the determinants, interventions, and outcomes agreed upon by stakeholders, the program rationale is an excellent base from which to conduct evaluations of various kinds.

STRATEGIES FOR
ARTICULATING PROGRAM RATIONALES

Evaluators invited to help stakeholders with a program rationale need to know that a range of evaluation strategies and approaches is available to them. Some options are *most* fruitful when used in certain limited circumstances, but it is important to realize that the strategies and approaches are not mutually exclusive. Evaluators are free to apply a combination of them to serve stakeholders' needs. Three strategies from Table 3.1 are explored in depth below: the background information strategy, development facilitation strategy, and troubleshooting strategy. The first is principally for designers and key stakeholders uncertain of the most pressing needs of a community or the kinds of intervention likely to be accepted by it. Using approaches such as needs assessment and formative research, evaluators systematically record the missing information. The second strategy, development facilitation, typically employs the working group or intensive interview approach to facilitate stakeholders' work on a program rationale. The last strategy, troubleshooting, may be helpful when stakeholders or evaluators believe the feasibility of an existing program rationale needs to be field tested. The plausibility testing approach can generate preliminary information about the program rationale's assumptions.

BACKGROUND INFORMATION
PROVISION STRATEGY AND APPROACHES

With the background information provision strategy, evaluators can gather pertinent empirical information about community needs, target group characteristics, and clients' and implementers' perspectives on interventions. This information helps program designers devise a program rationale. The evaluation approaches called *needs assessment* and *formative research* are suited to the collection of background information for stakeholders. Both are well-known evaluation approaches, and the terms have come to be used interchangeably. Despite their similarities, though, each approach has its distinct focus. Because needs assessment is especially suited to identification of unmet needs in a community, stakeholders puzzling out the goals of a program will be well served by it. *Formative research* is better at empirically describing clients' cultural backgrounds and capabilities; it is also useful for pinning down clients' (and implementers') opinions about proposed programs. This is an important distinction.

Needs Assessment

Needs assessment consists of research procedures to identify, measure, and prioritize the needs found in a community; it has been discussed intensively in the literature (e.g., Rossi et al., 2004; Witkin & Altschuld, 1995). The needs assessment strategy can facilitate selection of goals and target groups. Unmet needs in a community can be identified from existing data (e.g., census findings, vital statistics, agency records) and/or by studying a population via surveys, focus groups, or interviews. For example, one youth agency brought in program evaluators to assess needs among African-American youths in a community (Chen & Mark, 1996). The needs assessment results were intended for use in developing new programs to serve this target group. The evaluators surveyed the youths and their parents and identified a set of needs, assigning each a priority. Help with schoolwork was first, drug abuse prevention second, parenthood education third, recreation fourth, and health promotion fifth. The information pertaining to the top priority, schoolwork, helped stakeholders plan a tutoring program for the youths.

Formative Research

Background information can also be produced for stakeholders by using the formative research approach, which looks further than a community's unmet needs. Formative research is a systematic method of gathering empirical information about potential clients and implementers as well as their views of proposed programs. The formative research approach investigates these individuals' characteristics and cultures, and samples their opinions about proposed goals and interventions drawn from program rationales and plans. Popular research methods associated with the formative research approach are the focus group, the interview, and the survey. Focus groups tend to conserve time and resources, as compared to the other research methods, and they often generate more innovative ideas. Interviews tend to produce data that are more detailed and more reliable than data from other research procedures. Finally, although surveys have the advantage of including comparatively large numbers of individuals within a representative sample, the information they can provide is relatively limited.

A good illustration of the use of formative research can be found in a program in Hawaii to teach primary school students about preventing skin cancer (Glantz, Carbone, & Song, 1999). Planners sought program evaluation to help them understand the background of their target group as they proceeded to refine their program rationale. They wanted to know more about the students' ideas and activities concerning sun protection and skin cancer; they saw the usefulness of comprehending the target group's likely response to various educational materials and sun safety policies. Group discussions, focus groups, and interviews were conducted with the children, their parents, and staff members from local recreation organizations. The data obtained indicated that sun protection practices were inconsistent, even though general awareness of skin cancer prevention was widespread. The research showed that the children did not understand what skin cancer was, and they did not perceive it as a threat. This information justified the program's goals. Furthermore, data from formative research helped these program planners choose their determinants and intervention strategies. It suggested that children were reluctant to don long pants, long sleeves, and wide-brimmed hats to protect themselves from the sun. Messages promoting this degree of covering up would appear extreme to the students, too extreme to accept. A better message, according to the formative research, would urge gradual change, beginning with wearing short sleeves, longer shorts, and caps. Research findings for the parents and recreation staff showed their concern, also, that changes be acceptable within the culture of their tropical home, where, for generations, the preferred mode of dress has been light. These adults were, however, supportive of education and policy aimed at improving their own children's sun protection habits. As to the type of materials and strategies that would reach the adult *and* the juvenile audiences, the formative research asserted the importance of creative, engaging sun protection messages, some delivered in recreation facilities, some to be shared at home. Thus, not only did the formative research help with development of a program rationale, it also provided input for program development generally. By the way, this example points up the difference between formative research and needs assessment. Needs assessment showed planners that better sun protection was actually needed by the target population, whereas the formative research conducted honed in on program features most likely to bring them closer to fulfilling the need.

THE CONCEPTUALIZATION FACILITATION APPROACH—A PART OF THE DEVELOPMENT FACILITATION STRATEGY

Program designers and key stakeholders may have plenty of ideas for their program rationale. However, they often do not know how to clarify their thoughts and connect them systematically and coherently. The evaluator who is asked to help them can turn to the *conceptualization facilitation approach.* In this approach, the evaluator becomes a facilitator of stakeholders' work to conceptualize their program rationale or plan. There are certain guidelines for effective application of the conceptualization facilitation approach.

Working Group or Intensive Interview Format?

To use a conceptualization facilitation approach, stakeholders and evaluators first need to select either the *working group* or *intensive interview* format (these were introduced in Chapter 3). The intensive interview format is typically reserved for small-scale programs in which a few key people are charged with developing a program. Evaluators interview these key individuals with one of two aims: to clarify their program rationale, or to review the conceptual framework of the change model and draft a program rationale. For large-scale programs, the working group format is particularly well suited because consensus among numbers of stakeholders is always a driving issue. As a strong consensus builder, the working group format can create support for a program rationale.

Theorizing Methods

Once a format is selected, the evaluator and stakeholders must determine if forward reasoning or backward reasoning is the better theorizing method for their purpose. Stakeholders' backgrounds and preferences play a role here. Some program designers and other stakeholders may have extensive experience and expertise in various interventions. If their minds are set on a favorite intervention, forward reasoning is the preferred method of developing a program rationale. Stakeholders adept at social marketing to promote health, for example, may decide to seek funds for such an intervention in their community. Using forward reasoning, the evaluator could facilitate these stakeholders' identification of determinants affected by the intervention and help them estimate the eventual outcome.

With different stakeholders—say, a group that remains open to a process of selection—backward reasoning may effect a successful search for the right intervention. Backward reasoning demonstrates great flexibility. Among other things, it can help identify the determinant a program should address. For example, a working group wanting to serve a community by decreasing summertime delinquency (the outcome) among disadvantaged youth (the target group) decides to begin its task by identifying a tractable cause (for some, a more familiar term is *leverage*) of the problem—that is, a determinant. The group may determine that the proper determinant is involvement of youth in legitimate social, recreational activities. It would next specify an intervention, such as summer camp, that is expected to activate the determinant, in turn alleviating the delinquency.

Four steps from the development facilitation strategy remain to be discussed. Backward reasoning will provide the basis for their explication.

Identifying the Problem. The first requirement of development facilitation is to identify the problem a program aims to address. This is a relatively easy assignment for members of the working group (or interviewees). Still, if needs assessment or formative research data are available before the scheduled working group meeting (or interviews), it can benefit stakeholders immensely with its systematic exploration of community needs. With such data at hand, their decision should at all events be an informed one.

Identifying a Target Population. The next requirement of development facilitation is to specify a target population or group of the program. Evaluators can foster a reasoned choice of target by asking the working group (or interviewees) to consider two things: the *requirements or preferences of the funding agency,* and *community needs* that may have been pinpointed by needs assessment. Funding agencies frequently note, in the application materials for offered grants, that specified populations must be served by a funded program. Obviously, agency-issued guidelines (general or specific) for naming the target group must be heeded.

Furthermore, it is very desirable that programs serve those people most in need, or at highest risk; this must be the top priority in most cases. Exactly who is neediest or most vulnerable, however, is likely to vary from community to community. With HIV transmission, for example, some communities' high-risk population is men who have sex with other men, whereas other communities' most vulnerable members would comprise people who inject drugs (or migrant workers, or the homeless, or sex workers). Those working group members with long experience working directly with clients are good sources of information about community groups that have unmet needs. (Most working groups include at least a few such members.) Applying needs assessment or formative research—or perhaps gleaning existing data, such as epidemiological reports or vital statistics—also generates such information. Occasionally, it happens that the working group or evaluator identifies a need that the funding agency is not likely to want to fill. In such circumstances, the working group should be forthright in telling the agency that the need has been identified, and it should try to work with the agency to resolve differences.

Identifying Final Goals and Measurable Outcomes. The next step in the development facilitation sequence is to finalize the program goals and establish measurable outcomes to prove their attainment, in light of the specified problem and target group. General directions, and perhaps a list of goals, have often come from a funding agency or foundation, appearing in the grant application or call for proposals. Also, stakeholders usually have firm ideas about goals that a program should pursue. In moving toward establishing program goals and outcomes, evaluators should foster discussion of several issues. The first is *desirability versus plausibility.* There is a tendency among stakeholders to formulate goals that reflect very high expectations, outstanding aims whose desirability cannot be contested. Although glorious statements of program goals can inspire people and help build coalitions, they can also be unrealistic. A program with goals that are desirable but not plausible is destined to fall short of the mark. From program evaluation viewpoint, programs need practical goals—goals that can be reached with the resources available and that have some connection to existing knowledge and experience. Program designers who wanted to reduce drug use in a community largely by distributing "Say No to Drugs" buttons, for instance, would find scarcely any existing data indicating that a button distribution program is leverage enough to curtail drug use. For this program, it would be unrealistic to establish a goal of reducing drug use in the community. The evaluator does a favor to stakeholders when he or she emphasizes how the statement of program goals must exhibit plausibility and practicability in terms of available resources, proposed interventions, and the nature of the community problem. For example, suppose a media campaign against racism is proposed that would center on statements, published in newspapers and broadcast by radio, iterating the need to work together. The stakeholders in this antiracism program might be inclined to say that the goal of their campaign is less racism. It is certainly a noble goal and attractive to many stakeholders and others, but racism is a deeply rooted, complicated problem and has been for many years. It is questionable, to say the least, that racism could be solved by a brief media campaign. Should this program ever be evaluated based on such a grand goal, it would have to be deemed a failure. To prevent this, the evaluator works with stakeholders to arrive at practical goals for the program: perhaps, in this case, enhancing people's awareness of the problem of racism, or building support for long-term strategies to reduce it.

Goals themselves can be long term or short term. Short-term goals are attainable in a few months to a year, and most can be achieved. Long-term goals are the ultimate aims for which a program strives. As *ultimate* goals, they typically require a great deal of time to achieve; furthermore, their achievement is contingent on many things. A short-term goal for a homeless program might be to shelter homeless people for a certain period. Its long-term goal could be to help these same individuals find work, attend job training, or obtain more permanent housing. Assisting homeless people in finding and keeping jobs is a much more challenging task, however, than locating a bed for the night. Many months could go into addressing job readiness alone because issues such as inadequate job skills or a poor work ethic will have an effect on the outcome. It would clearly take a long time for a program to manage these issues and achieve the long-term goals.

And yet long-term goals are very important to stakeholders in many situations; such goals are the reasons an organization or agency is entrusted with funds. The point is that the work group must identify short-term goals as well as long-term ones. Short-term goals easily provide the "measuring sticks" with which program staff can see their successes and be motivated to press on with their work. The evaluator should try to see that a working group finalizes both short-term and long-term goals. The desirability/plausibility and short-term/long-term issues have a certain relationship. To attract funding, program stakeholders may believe it necessary to set extremely desirable—but simultaneously unrealistic and impracticable—goals. Indeed, it may be possible for a working group to embrace such goals in the long term, while setting more practical goals in the short term. In the example of the media campaign, for instance, reducing racism could indeed be the long-term goal of the messages, while enhancing awareness of racism and the supports available to deal with racism could be short-term goals.

An evaluator participating in a working group may soon find that stakeholders sometimes confuse goals and objectives with action steps. The evaluator should make the differences clear. Action steps are activities that create or strengthen an implementation system. Action steps, then, relate to the elements of the action model. Examples of action steps are "to hire three additional outreach workers and a program coordinator," or "to build a management information system for fiscal management." Objectives are the achievements of the implementation on its way to reaching its final goals. In many cases, objectives are given in terms of action steps, meaning that statements such as "to hire three additional clinicians in the first quarter of the project" or "to provide treatment to 50% of clients in three months" should not be regarded as program goals. In addition, the evaluator may need to reiterate to stakeholders the distinction between goals and outcomes. Ideal statements of program goals are concrete and concise, indicating the program's purpose and conveying key stakeholders' interests and concerns. Most, however, speak generally and in the abstract. This fact makes it even more necessary for the working group to go further, not stopping with statements of program goals but also—based upon those statements—finding a way to express the goals in clearly defined and measurable terms, which are called *outcomes.*

Clearly defined program outcomes are fundamental to the operation of any program. Outcomes provide tangible yardsticks for measuring accountability variables as required by funding agencies or the public. Could a substance abuse program satisfy those who hold it accountable by telling them that the program really did assist drug abusers in reaching a meaningful reduction in drug use and achieving social and economic functioning? Probably not, because these terms are not measurable. What exactly do the words "meaningful reduction" mean? Is cutting back from drug use three times daily to twice daily "meaningful"? Similarly, "social and economic functioning" is not a precise, measurable concept. To *matter* when it is time to get down to accountability, a program goal must be rendered measurable—in which case it has become an outcome.

Consistency among program goals can be as important as clarity. An intervention program often has multiple goals. One task for the working group is to ensure that these goals (and corresponding outcomes) are not incompatible. The program evaluator is prepared to point out to stakeholders any goals or outcomes that are inconsistent or even mutually exclusive. For example, suppose that a family court proclaims two goals for its handling of child abuse cases: providing abused children with security and support, and keeping families intact. At times, these two goals may simply be incompatible. Pursuing the goal of keeping families intact might require returning children to abusers, compromising their safety. Conversely, pursuing the goal of providing security and support to children might require sheltering them apart from their biological parents, in which case the family is no longer intact.

Identifying Determinants Likely Able to Change Outcomes. Tied into choosing achievable goals is understanding the determinants at work in the problem being addressed. Determinants are leverages or forces—the causes of a problem believed to be linked directly to the production of outcomes. Some stakeholders may substitute terms like "intermediate outcomes," "root of the problem," "mediator," "social and psychological facilitator," or "social and psychological barrier" for "determinant." Evaluators need to figure out which terminology is being used by stakeholders and communicate with them accordingly. Once they are activated or changed, determinants can change outcomes. Evaluators can show their working groups which determinants must be their focus if a program is to attain its goals; this forms the next step in completing the program rationale. The causes of problems can be many. Spouse abuse can arise out of husbands' low self-esteem, wives' lack of economic power, husbands' responses to stress, the patriarchal society, husbands' ignorance of the law regarding spouse abuse, or any number of other complexities. Which of these determinants can an intervention program address? Obtaining a content expert or performing a literature review can educate working group members about leverages that have potential to create change. Because a program is usually constrained by its resources, the working group needs to identify one, or at most a few, major determinants in line with the mission and expertise of the implementing organization. It might be appropriate, to continue with our example, for an intervention program against spouse abuse to limit itself to the determinants of low self-esteem and unskilled response to stress. Selection of determinants has a direct bearing on selection of interventions to be employed in a program. A program to undo an increase in youth crime will be able to address only some of the potential factors in it: insufficient parental supervision, peer pressure, poor school performance, drug abuse, inadequate recreation opportunities, child abuse, and lack of positive role models. Perhaps designers of the program decide to focus on one cause, peer pressure, as the determinant. Their choice means that interventions built into the program will center on peer pressure.

Determinants and Types of Program Theories

A program's determinants are identified according to the type of program theory on which the program is based. Programs based on scientific theory (see Chapter 2) can usually find within that theory some statement of appropriate determinants. For example, a program built on the theory of planned behavior (Ajzen & Fishbein, 1980) will draw from that theory a wealth of information about *intention to act* as a fundamental determinant in behavioral change. Programs built on stakeholder theory, by contrast, receive no such guidance, leaving stakeholders responsible for examining the assumptions underlying their choice of determinants. The evaluator can help clarify the stakeholder theory, if necessary, making implicit assumptions explicit. Steps in this facilitating process include providing examples of determinants to acquaint stakeholders with the concept; asking stakeholders to name the determinants they believe will most affect program outcomes; listing these major determinants; evaluating these determinants in light of time and resource constraints on the program; and, finally, asking stakeholders which determinants they desire to *and can afford to* select as their program's major focus. When it has been particularly difficult for interviewees or working group members to latch on to the concept of the determinant, the evaluator can try substituting the terms "intermediate outcomes," "causes of the problems," "social and psychological barrier," or "social and psychological facilitator" for "determinant." Then, probing questions from the evaluator might take such forms as "Can you name intermediate outcomes that the intervention needs to reach first before attaining the ultimate outcomes?" "Can you name crucial social or psychological barriers preventing clients from achieving program goals? (Think clients' social and psychological barriers that the program intends to remove.)" and "Can you name crucial social and psychological facilitators helpful to clients who are trying to achieve program goals? (Think clients' social and psychological facilitators that the program intends to restore.)" The listing of potential facilitators and barriers that should emerge will be the equivalent of the listing of major determinants cited earlier in this paragraph.

An occurrence in Taiwan offers one effective illustration of the dynamic when a working group identifies the cause of a problem. Such a group was formed there in the 1980s to respond to increased suicides by police officers during a time of transition from a one-party political system to a multiparty political system. In a meeting, some members of the group, including psychologists, argued that these suicides were mainly attributable to police officers' lack of stress management skills. Stress was intensifying because mass demonstrations had continued in the wake of abolishment of martial law, and these members advocated conceptualizing the suicides as a problem of individual adjustment. Remaining members of the working group, however, including sociologists and program evaluators, suggested that there were other influences, one of which was the sudden lack of social norms following the demise of martial law. Under martial law, there had been no salient structural distinctions between the ruling political party—the Komington—and the government. The Komington was the government, and vice versa. Police officers belonged to the Komington's law enforcement arm; their authority was legitimate. But that authority weakened as martial law crumbled, and the public now regarded police officers as puppets of the Komington. The police were ridiculed and humiliated in public and in the media. This sudden identity crisis might have something to do with the increase in suicide by officers, the second faction countered. The decision eventually made by this working group—in its entirety—was to place their program on the individual level rather than the structural level. By identifying personal turmoil as the determinant, the working group felt that it would subsequently be rather easy to design and implement a counseling program for police officers. The working group also, however, notified decision makers of the structural problem and recommended that they address it. This exchange among stakeholders and evaluators in Taiwan shows how alternative views can be brought to stakeholders' attention with good result. At the same time, it demonstrates the value of barring evaluators from pursuing personal agendas or substituting their own values for stakeholders' values. In short, an evaluator's role is to ensure that stakeholders make *broadly* informed decisions.

Choosing Interventions/Treatments that Affect the Determinant

When determinants have been identified adequately for the purposes of the program, the evaluator's work shifts to facilitating the working group's selection of an intervention that can activate those determinants. There are usually a number of intervention options. For example, if a program rationale posits that truancy is a major determinant of burgeoning youth crime, and if truancy therefore will receive the program's focus, there are many ways to try to change this determinant. Schools might provide counseling to help students with academic or other school difficulties. They might develop new curricula to appeal especially to troubled youths. A policy might be enacted that fines parents when their children are absent; sends school administrators to immediately visit any student who has not shown up for class; or authorizes police to write citations to students not in school during school hours, requiring both the student and parent to appear in court. Pinning down the best intervention involves four criteria that the evaluator should share with other members of the working group:

1. *Mission and Philosophy of the Organization.* First of all, the intervention selected by the working group must be appropriate to the implementing organization's mission and philosophy. Conflict between the intervention strategy and the implementing organization's values creates too much stress and can interfere with implementation of the program.

2. *Budget and Personnel Restraints.* Second, the chosen intervention must reasonably reflect the budget and expertise of the implementing organization. No organization can adopt an intervention far beyond its means, nor can it agree to an intervention requiring personnel it does not have.

3. *Theoretical Justification.* Third, an intervention needs sound theoretical justification in order to be effective. After all, theoretical justification is scant or weak for good reason when it *is* scant or weak. As we have seen, a program founded on scientific theory finds well-reasoned determinants in that theoretical ground, and stakeholder theory-based programs can justify a

choice of determinants by citing other programs, literature, or common sense. It is possible to justify a job training program (which is an intervention) by noting that it enhances clients' job skills (which is the determinant) and by noting further that better job skills can lead to employment or better paid employment (which is the program goal). It is possible to do this because the relationships among job training, acquired job skills, and employment comprise a well-recognized common experience that makes sense to most of us. It is less possible to convincingly justify "say no" buttons (an intervention) as an influence on perceptions of drug use (the determinant) capable of discouraging drug use in a community (the program goal) because our experience tells us the intervention is too weak to produce such a characteristically hard-won change.

4. *Base of Evidence.* Fourth and finally, other things being equal, an intervention strategy that is supported by empirical research should be preferred to intervention strategies without such support. This principle is not meant to discourage innovation, but the existence of empirical evidence, even indirectly linked empirical evidence, does provide additional confidence in an intervention.

THE RELEVANCY TESTING APPROACH—A PART OF THE TROUBLESHOOTING STRATEGY

In some situations, program designers have completed a program rationale but wonder whether the proposed change model it includes is relevant to the problem as it is experienced by target group members in their world. A program evaluator might be called on to carry out empirical checks—field testing—on the program rationale before deploying it to create a program plan. The *relevancy testing approach* can meet the needs of program designers in this position. Relevancy testing comprises a reality check. It is a rapid appraisal, from the field, of the fitness between the problem faced by the target groups and the selection of goals, determinants, and interventions for a program. It is a very useful approach for catching potential weaknesses and improving the quality of a program rationale. Stakeholders participating in relevancy testing are most often the implementers and clients of the program. Like formative research, relevancy testing is flexible as to the research methods involved (focus groups, interviews, and surveys are options). The program evaluator conducting relevancy testing is likely to find three questions very important to her or his progress:

1. *Are the goals stated in the program rationale appropriate and reasonably relevant to the identified problem and target group?* In other words, is it realistic to envision the targeted individuals achieving the established goals? Sometimes, the original program rationale includes goals beyond the scope of the program's interventions or ill-fitted to clients' problems and needs. Weeding out such inconsistencies begins with asking clients and implementers simply to comment on the program rationale. As an example, a perinatal care program prepared a program rationale proposing to employ a midwife to care for 300 pregnant women, and deliver their babies, in one year. The evaluator brought together a group of midwives and asked for comments. The group's immediate response, based on its extensive experience, was that meeting this goal was impossible. Without resources above and beyond those available to the program, the group felt no more than 100 women could be served in a year.

2. *Does the program rationale name a determinant likely to have an impact on the target population?* A program rationale assumes that a particular determinant causes a problem affecting a target population, and it proposes an intervention to alter the determinant, creating desirable outcomes. Those on a program's front lines—people such as clients and implementers—can help the evaluator understand whether a determinant vital to the program rationale is sufficiently relevant to the proposed target population. Teen pregnancy prevention programs provide an example. Such programs may assume that adolescent girls become pregnant because they are unskilled in using birth control devices. Making that assumption, such programs set out to teach the girls about the devices, both to make them comfortable with using them and to prevent frustrating, unsuccessful attempts at using them. The program's two determinants are, then, lack of skill and frustration. To test the accuracy of these determinants, the evaluator invites teenage girls to participate in focus groups. If discussions suggest that the girls in fact know a lot about using birth control, the evaluator will tell program planners that the selected determinants may not be appropriate for this target population.

3. *Will the outlined intervention be reasonably acceptable to the target group?* An intervention that is well received by one group can be offensive, or completely irrelevant, to another. The evaluator engaged to field test a program rationale must ensure that the proposed intervention or treatment will be

both relevant and inoffensive to the particular population being targeted. One major issue is cultural sensitivity. Some societies are more conservative than most Western societies. A lesson on birth control that includes the demonstration of condom use before a group of young women will not be helpful if the audience finds this insulting or painfully embarrassing. Making inquiries about the cultural sensitivity of proposed interventions is one task for the program evaluator. Another is determining whether the messages implicit or explicit in an intervention will be comprehended by clients. This determination involves both clients and implementers. Synthesizing the comments of these two parties, the evaluator learns whether the language of the message is understood by clients and is respectful of their culture; *and* whether implementers feel confident that they can effectively communicate to clients the curriculum within the intervention.

Research Example of Relevancy Testing

In Chapter 2, a home-based intervention program to reduce passive smoking by infants (Strecher et al., 1989) was used as an example, and we return to it here to illustrate relevancy testing. This program's rationale can be represented graphically as in Figure 4.2.

Strecher and colleagues (1989) tested the relevancy of this rationale in the field by recruiting 104 mothers of infants (40 recruited at clinics in face-to-face interviews and 64 contacted for a telephone survey via county birth records available to the researchers). They obtained information that answered the three questions above and clarified assumptions underlying the program rationale. First, for example, they asked whether the program's stated goals were relevant to the target population. The notion of urging mothers to stop smoking altogether had been brought up by program planners, but the researchers wondered whether a more modest aim (providing a smoke-free environment for the infants) was more realistic. They queried the mothers about their interest in giving up cigarettes; the resulting data showed that most would not be inclined to try to quit smoking during this period in which they already faced the many stresses of caring for a new baby.

The researchers also asked whether the named determinants, outcome expectation and efficacy expectation, were likely to make an impact on the mothers. Outcome expectation comprises one's belief about whether a given behavior will lead to a given outcome; efficacy evaluation comprises one's belief in one's own capability to perform the behavior that leads to the outcome. In the example, the

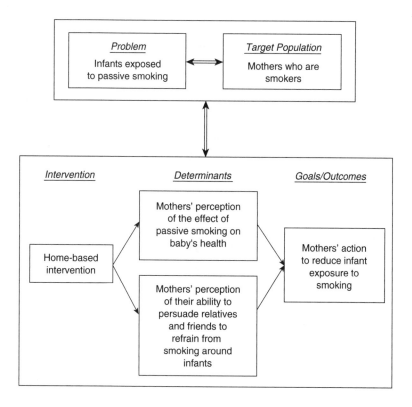

Figure 4.2 Program Rationale of a Program to Reduce Infants' Exposure to Smoking

SOURCE: Adapted from Strecher et al. (1989).

obtained data showed that mothers had general knowledge of the effects on babies of active smoking by adults: respiratory problems. However, according to the findings, many mothers did not understand the actual nature of passive smoking. For example, they believed that a baby inhaled significant amounts of cigarette smoke only when the smoke was blown directly into the baby's face. Few realized that significant amounts of smoke are also inhaled by a baby when it is across a room from a burning cigarette or when it enters rooms where people had been smoking earlier. Findings like these informed the researchers about the outcome expectation determinant.

Other findings informed them about the efficacy expectation determinant. In general, the researchers found low levels of perceived efficacy among the new mothers, reflecting a lack of confidence in their ability to persuade husbands, other relatives, friends, and caregivers to refrain from smoking around the infants. Strecher et al.'s relevancy test prompted fine-tuning of the program rationale in the following areas: An emphasis was placed on teaching mothers what passive smoking actually is; the intervention was refocused on the immediate, specific outcomes of an infant's exposure to secondhand cigarette smoke; and the program aim was reoriented toward helping mothers maintain smoke-free environments for their babies. The issue of smoking cessation was discussed only if a mother or other family member expressed interested in quitting.

Moving From Program Rationale to Program Plan

In this chapter, the importance of the program rationale has been demonstrated, including its position as the base for the developing program plan. In addition, we have reviewed ways to facilitate stakeholders' development of their program rationales. Once a sound program rationale has been adopted, stakeholders proceed to devising the program plan, and program evaluation is a valuable facilitator of this work as well. This is the topic of Chapter 5.

≠ FIVE ⱶ

HOW EVALUATORS ASSIST STAKEHOLDERS IN DEVELOPING PROGRAM PLANS

———•◦•———

Program evaluators are asked to help in developing program plans just as often as they are asked to help formulate program rationales. In this chapter, program planning strategies and techniques are presented in terms apropos to start-up intervention programs. However, the discussion may also prove useful for established programs whose stakeholders decide to fine-tune or revise an existing program plan. As Chapter 4 sought to make clear, a sound program plan is one that has been guided by a program rationale. So, again, the evaluator asked to facilitate production of a program plan in the absence of a program rationale will need to approach the stakeholders about the importance of adopting a rationale before moving on to a program plan.

THE ACTION MODEL
FRAMEWORK AND THE PROGRAM PLAN

Program plans are blueprints for the actions prescribed by program rationales. Early on, program plans guide the organization of program activities and the allocation of resources. Later, they stipulate the program staff's day-to-day operations and coordinate disparate personnel units. The quality of a program's plan affects the quality of its implementation and, eventually, the

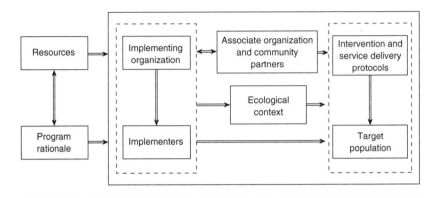

Figure 5.1   Conceptual Framework of a Program Plan

degree of its effectiveness. High-quality program plans come most readily from clear and realistic program rationales. A program rationale that never quite says how to reach and screen the target group of the program complicates the work of program managers trying to determine the kind of implementers to recruit and the kind of training needed to prepare to deliver services.

Just as the program rationale belongs to a program's stakeholders, the program plan is theirs, too. It is the stakeholders' duty to develop the plan for their program; but when the complexity of a program requires it, stakeholders seek help from experts like program evaluators to ensure a program plan that is sound and feasible. An evaluator invited to facilitate the program plan's development has a resource in her or his general knowledge of action models (which is part of program theory). The number of activities called for in most programs can be dizzying, even to stakeholders. They pose a challenge to evaluators as well, who may need to help stakeholders conceptualize all of the various activities within a meaningful scheme in order to manage implementation successfully. As a guide to that scheme, the book offers in Chapter 5 a general conceptual framework of an action model, the *action model framework*. This framework, which is represented in Figure 5.1, is a conceptualization of a generic program plan, useful in developing any number of specific, situational program plans.

At the top of the block in this figure sits the program rationale. Its role is to direct the development of the program plan, which includes six components, often in a standard sequential order; that is, certain components must be devised before others can be (as shown in Figure 5.1). Implementing a program plan requires a capable, committed implementing organization. It is up to the implementing organization to find and train responsible implementers, who must have ample skill and commitment. The implementing organization (and its implementers) also must connect to associate organizations and community partners so that services may be delivered efficiently and must generate interpersonal and community support for the program. Clients from the target group must be reached and motivated to join the program, which is another job for the implementing organization. Only with all of these components in place can implementation of the intervention protocol be launched and services delivered to clients. In spite of their sequential order, all six components are interconnected. When new information alters a component, that tends to lead to change in the other components, as when stakeholders decide to include additional target groups in the program. This move immediately affects the implementing organization's need to coordinate activities, the intervention activities themselves, the content of implementation protocols, and other components as well. Examination of Figure 5.1 also shows that the relationship between program rationale and program plan need not always be unidirectional. On occasion, in the process of developing the program plan, stakeholders may become aware of problems or weaknesses in the program rationale, revising it to correct these. Figure 5.1 illustrates this kind of feedback process within the action model framework.

The action model framework is of help whether the evaluator is facilitating (a) the development of a new program plan, (b) the clarification and strengthening of an existing program plan, or (c) communication about the program plan. It is also useful later on in the process in terms of (d) preparation for the formal evaluation of the program's implementation.

The New Program Plan. Program designers initiating a program plan depend on input from program directors, implementers, and other key stakeholders. The action model framework is a means of systematizing the distinct pieces of this collective effort. It helps to ensure that no important issue is left out and that no gaps are permitted to compromise the quality of the program plan.

The Existing Program Plan. When stakeholders have come to wonder about the completeness or soundness of a program plan they have devised, the action model framework makes a good tool for evaluators to use to get stakeholders working together again. This purpose for the framework is at times very similar to the next, facilitating communication.

Communicating About the Program Plan. The action model framework shows the way each of many program activities ties in with others. Larger programs may include hosts of activities, which the framework allows to be ordered into categories, lending meaning to the incipient program. As stakeholders begin to see the proverbial "method to the madness," they become better able to discuss the program plan with each other and their various constituents.

Planning Ahead. It is generally a given that, once under way, a program will need to be formally evaluated. The commonly ordered *process evaluation* is research into how closely the actual implementation of a program matched the stakeholders' intentions. The action model framework can make clear precisely what their intentions were, a valuable aid to the evaluator designing and conducting a process evaluation. Similarly, when outcome evaluation is requested by a program, the action model framework can answer the evaluator's questions about how interventions were to be delivered or how the target group was to be reached. Knowledge of such information is essential for designing an outcome evaluation.

STRATEGIES FOR DEVELOPING PROGRAM PLANS

Calling on a program evaluator to assist in preparing both the program ratio-nale and the program plan—as stakeholders, we read in Chapter 4, frequently do—is certainly advantageous. Evaluators have the knowledge and skills to note consistency and to ensure that all crucial elements of a program plan are securely in place. Furthermore, the evaluation strategies and approaches for developing program plans meld with those for drafting program rationales, and preparation of the two is, for the practical program evaluator, essentially a unified project.

Certain strategies and approaches for assisting stakeholders with program plans are presented here, with examples of their application following. They are not mutually exclusive strategies and approaches; it is sometimes benefi-cial to include more than one tactic during development of the plan. The back-ground information strategy can provide stakeholders with missing general information about communities and clients to get their program plans under way. The formative research evaluation approach fits well for this purpose. The development facilitation strategy can be ideal to use with stakeholders who feel they need evaluators' input to develop the plan for their program, or evaluators' support to build consensus on the worth of a program plan. The troubleshooting strategy comprises small-scale trials of the program plan to iron out potential problems and come up with ways to fine-tune the plan. Selecting among, and then using, these strategies and approaches can be complex; guidelines for the process appear in Chapter 3.

THE FORMATIVE RESEARCH APPROACH, UNDER BACKGROUND INFORMATION PROVISION STRATEGY

Chapter 4 illustrated how the formative research approach generates meaningful background information for stakeholders working on a program rationale. The same principles and procedures of formative research are valuable tools for assisting stakeholders at work on a program plan. Simply put, the evaluator will find it very feasible to design and conduct *generic* formative research incorporating background information useful in composing both the program rationale *and* the program plan. However, certain matters are especially pertinent when working with stakeholders on a program plan. These are numbered below; see Chapter 4 for the more general principles and procedures of formative research.

1. *Formulating Research Questions for Informing the Program Plan.* The action model framework can play a role in determining which questions should be researched to equip program designers or working groups with adequate understanding to write a program plan. Examples of research questions are, "What factors discourage target group members from participating in the program?" "What components should be included in the intervention protocols?" "What mode of service delivery will be acceptable to these clients?" "Does the proposed implementing organization have the needed capacities?" and "What training will implementers require?"

2. *Gathering Data to Answer the Research Questions.* Flexible research methods are a hallmark of the formative research approach. The crux is returning feedback to the program designers or working group *quickly.* Focus groups, intensive interviews, and surveys all can provide good information. Research participants are typically prospective clients, implementers, and other stakeholders.

Example of Formative Research

Formative research was the mode selected by Gettleman and Winkleby (2000), who set out to learn what might be the best structures and implementation schemes for programs addressing incipient cardiovascular disease (CVD). Low-income women—African-American, Hispanic, and White—constituted one of the largest populations at high risk for CVD, but insufficient information existed on how to reach them in meaningful numbers. Gettleman and Winkleby's research involved seven focus groups with 51 low-income women in their communities. The focus groups showed that these women preferred receiving health information in "visual" formats as opposed to text-only formats. They also felt most positive about prevention programs that addressed several risk factors, particularly smoking, lack of exercise, and high-fat diets. Testimonials from healthy women who described how they embraced heart-healthy behavior, and factual commentaries from physicians, were two more elements of a prevention program the focus groups thought effective. The CVD intervention emphasized staying healthy for one's own sake, as well as heart-healthy behaviors and skills, and the role of choice in behavior change. Formative research during this project also asked the women about barriers to and incentives for participation. Three identified barriers were a lack of time, transportation, and child care; most said they would not participate in a CVD intervention program unless there was free child care. As to constraints on time and location, the women suggested intervention programs be held at job sites during lunch hours, for those employed outside the home; and at community sites, for others. The researchers concluded that transportation problems could be eased by offering program activities in places women routinely go with their children, such as public libraries or pediatric clinics. Child care, free meals, and cash or food vouchers were incentives the research identified as holding the potential to encourage participation. All of this information proved valuable in developing interventions tailored to low-income women.

THE CONCEPTUALIZATION
FACILITATION APPROACH, UNDER
DEVELOPMENT FACILITATION STRATEGY

As has been illustrated in Table 3.1, within the development facilitation strategy is an evaluation approach called *conceptualization facilitation,* one that can serve very well to help stakeholders with their program plans. Although we treat conceptualization facilitation as a distinct evaluation approach, using it to develop a program plan involves the same principles and procedures that Chapter 4 presented for helping stakeholders develop program rationales. For instance, the communication formats available for developing rationales (intensive interview, working group) are also used frequently in program plan development, with just a little modification. What may be changed is the membership of the working group or the people to be interviewed. Such modification ensures that both administrators and implementers are well represented during development of the plan; each party brings expertise to the work and holds high stakes in the planning of the program. Working from the program rationale, the evaluator can discuss thoroughly with stakeholders the elements of the action model framework and how these relate to a program plan.

Once more, the elements, or components, of the action model framework are the target population, the intervention and implementation protocols, the program implementers, the implementing organization, peer organizations and community partners, and the ecological context. The evaluator's first task when applying the conceptualization facilitation approach is to assess the importance of these six components to the proposed program. Four of the components are considered crucial in designing any program plan; the importance of two others (peers/partners and ecological context) varies from program to program. In general, larger programs should assign relatively more weight to an implementing organization's capacity and to establishing linkages with peer organizations. Furthermore, a program acknowledged to be less than appealing to its prospective neighbors—such as a homeless shelter or halfway house—should give added weight to contextual supports for the program's implementation. When stakeholders are confident that a given component is irrelevant to their program, it may safely be excluded from the program plan. However, the reason for such a decision should be documented and all stakeholders made familiar with it.

While helping stakeholders with a program plan, program evaluators typically work to endow the six components of the action model framework with certain qualities. These have been touched on earlier in the book; what follows below is a step-by-step description of the six-part process of helping stakeholders produce an effective program plan. Evaluators can apply a working group or intensive interview format to assist stakeholders in developing a program. The following are discussed in the working group format. Readers can easily apply the same principles and tactics to the intensive interview format.

1. Implementing Organization:
Assess, Enhance, and Ensure Its Capacity

To select an implementing organization and, in the process, to measure that organization's capacity-building needs, a working group should consider three main factors: *technical expertise, cultural competence,* and *manpower and other resources.*

Technical Expertise. The working group needs to discuss what technical expertise potential implementing organizations need to have. Which organization is best equipped to deliver a particular intervention? In cases in which a group of implementing organizations has been selected prior to the existence of the working group, the working group can then make a recommendation of criteria for program directors and other key decision makers to examine the technical proficiency of the organizations to ensure their readiness to offer interventions.

Cultural Competence. The evaluators can bring the working group's attention to specifying what kind of cultural background and experience the implementing organizations need to have in order to recruit and train the implementers to communicate well with their clients or to be trusted by the target population. The working group can also indicate what kind of training the implementing organizations need to provide to develop and ensure the cultural competence of their implementers.

Manpower and Other Resources. The working group needs to work out a plan to select the implementing organizations that can strongly commit manpower and resources to the program, or the program probably will not be successful. Implementing organizations that are overloaded with existing projects should not be asked to implement the new program. This often happens within government agencies that can be assigned to start new programs by administrative order. The implementing organization often receives the order with no accompanying budget or staff increase. Busy with its existing duties—indeed, just as likely to be *overloaded* with them—a staff may feel little incentive to take on something else.

2. Intervention and Service Delivery Protocols:
Delineate Service Content and Delivery Procedures

The general nature of the interventions to be conducted by a program is laid out in the program rationale. In developing a program plan, however, the working group needs to go much further. First, it should specify in detailed terms the services to be provided by the program—the program's *intervention protocol*. Next, it must explain just as thoroughly the procedures involved in, and also the setting for, the delivery of these services—the program's *service delivery protocol*. No intervention can be carried out precisely in the field without complete intervention and service delivery protocols. An HIV prevention program provides an example of developing protocols. The intervention protocol is a description of the content, curriculum, intensity, and duration of the intervention services or activities to be provided to the target group. The working group should strive to note every detail. For instance, the HIV prevention program's working group might specify that a group counseling intervention take place over three weekly sessions, each 2 hours long. The working group would then elaborate on these sessions. For example, the first session should be an "icebreaker" for freeing clients of some of the reluctance they might feel about speaking openly of HIV risk; the second should be a discussion of barriers to safer sex practices and how to remove these barriers; and the third should teach information and skills for safer sex. The working group must arrange for the creation of a curriculum for each session in order to provide implementers with clear guidance on the topics and activities of each session; and also with any available advice for making each discussion session a success. Detailing the where and how of service delivery is the service delivery protocol on which the working group must also elaborate. An evaluator facilitating the work of a program designer or group drawing up protocols should be certain these individuals are familiar with potential modes of service delivery and potential service settings. *Modes of delivery* include categories such as the following:

- *One-to-one interaction*—an intervention delivered by an implementer to an individual client (one client at a time), such as individual counseling; labor-intensive and therefore costly, but also possibly one of the most effective modes of delivery
- *Support group*—an intervention via a group process of mutual understanding and support, such as Alcoholics Anonymous; group of several

clients plus therapist/facilitator meets regularly, opening minds to notion of change, mutually encouraging and accepting change, facilitating change, and sustaining change (one-to-one interaction and support groups are, of course, popular modes of delivering treatments as well as interventions)

- *Intervention classroom*—information or demonstration (e.g., a brief exercise routine) delivered to target group members by a presenter, often followed by question/answer time; inexpensive to provide but difficult to individualize
- *Documents/literature*—published intervention message in brochure, pamphlet, flier, or similar document, mailed to target group members or distributed for pick-up in public areas; potential to reach a range of people at relatively low cost but danger of information-weary readers ignoring or taking lightly the published message
- *Telephoning/Web posting*—calls from implementers to target group members to deliver an intervention, or intervention messages posted on Web sites thought to be frequented by target group members; constrained by clients' and implementers' access to technology
- *Mass media*—conveying of intervention message via television and radio, newspapers and magazines, hotlines, and so on; passive delivery method with no assurance target group will encounter message

An implementation protocol must also specify the desired service setting. Informed decisions can be made only when the working group or program designer is aware of the range of possible intervention settings. These include the following:

- *Office/clinic/hospital,* often belonging to the implementing organization and under its control (clinic/hospital typically used to deliver treatments), which is a logistical advantage *if* target group members have transportation to the facility; professional-type surrounding may impress clients or may feel too formal for comfortable disclosure of personal problems
- *School/community center/club facility,* often conveniently located, but facility directors can balk at providing space for use by certain populations (e.g., drug users, the homeless)
- *Public areas*—streets, parks, playgrounds, and so on—open to all; popular with outreach workers for initial contacts but services not

commonly deliverable on the spot; slim chance of more than brief interaction with busy people (going somewhere, supervising children, engaged in sport or exercise); time sufficient only for information sharing, not real intervention

- *Store/shop interior,* such as laundries, bars, salons, pool halls, drugstores, bookstores, and so on; popular with outreach workers seeking target group members; target attention may be focused better than in outdoor areas; proprietor's approval required, raising issue of feared impact on business
- *Private home* of implementer, volunteer, or client, when willing; intervention can be complemented by serving of food and beverage creating cheerful, casual atmosphere in which sense of security and relaxation prompt open sharing of experiences

When the program being planned is of the extended, labor-intensive variety—a mental health or alcoholism treatment program, for example—clients typically complete a series of stages, perhaps in various settings. The implementation protocol then is required to specify exact procedures for moving clients from one stage to the next: intake, screening, assessment, referral, treatment, revisit, and finally exit. Among these procedures should be safeguards preventing a client from "falling between the cracks" along the way.

3. Program Implementers: Recruit,
Train, and Maintain Competency and Commitment

Program implementers—the people who deliver intervention services—can be professionals or volunteers. As a working group or program designer explores who will serve as implementers, the program evaluator can assist by providing information on means to ensure the quality of their work. Stakeholders and the evaluator may also need to concern themselves with raising implementers' levels of technical and cultural competence; determining incentives to encourage implementers' commitment; and communicating clear direction, ample instructions, and firm expectations to implementers about their work. Programs planning to employ professionals will need the best available. Highly trained individuals are required for the most intensive treatments and interventions, such as certified teachers for education programs, therapists for drug abuse counseling, and social workers for case management.

In programs like these, compromising on the qualifications, experience, and commitment of personnel directly affects outcomes. Unqualified volunteers should never be substituted for professionals when providing services that require professional training. But volunteers can accomplish wonders for some programs—and not just in terms of saving costs. There is a tendency to use volunteers who have ethnic, social, and economic backgrounds similar to those of a program's target population. They may be most capable of reaching hard-to-reach clients, delivering information to them and escorting them to the intervention setting.

Quality assurance for the work of all implementers is achieved using any of three general strategies, according to the preference of the working group or program designer. The evaluator can facilitate decision making in this area as well. The three strategies are *training, technical assistance,* and *review.* Training provided by the implementing organization can establish or enhance implementers' skills and cultural competence, enabling them to deliver services effectively. Even well-prepared implementers, however, will occasionally meet with difficulty delivering services. For those times, the implementing organization needs some mechanism in place to assist implementers, particularly when the program being implemented is groundbreaking or very large. Finally, a crucial part of quality assurance comes from supervisors' periodic review of implementers' work. At times, for fairly obvious reasons, implementers may hesitate to share problems and mistakes with supervisors. If this becomes a concern within a program, perhaps the review of work can be conducted by peers. In fact, peer review often prompts valuable sharing of experiences in handling problems, frequently revealing innovative means for addressing them.

4. Associate Organizations/Community
Partners: Establish Collaborations

An implementing organization may need to create meaningful working relationships with various types of peer organizations. It is up to the working group or program designer to identify both the organizations of interest *and* strategies that are likely to launch such relationships. (This step in creating the program plan is, as previously noted, more important to the success of large-scale programs than small-scale ones.) The evaluator should lead the working group in a consideration of the four main types of associate organizations able to partner with the implementing organization. These are *core organizations, related organizations, auxiliary service organizations,* and *collaborating organizations. Core organizations* are those with which the implementing organization must work—very closely and efficiently—if its program is to be implemented. An education program targeting teenage drop-outs obviously needs a strong working relationship with schools, for the schools know both who has dropped out and, often, the reasons why. Furthermore, for many intervention programs aimed at minors, parents must be considered a core organization, in that their formal consent is required before their children can legally participate. For example, sex education programs for high school students typically require each enrollee to have a parental consent form on file. *Related organizations* are organizations that have the power to inadvertently interfere with program implementation if they are unaware a new program is under way. A program whose outreach workers will frequent districts known for drug trafficking or the sex trade would be wise to contact police departments in those areas first. Simple notification of this sort has saved many a program from difficulty. Sometimes, a new program must actually obtain a related organization's permission to begin implementation. *Auxiliary service organizations* are those that can meet clients' needs beyond those the implementing organization fulfills. For example, clients in drug treatment programs may also need shelter, food, work, and education. A good referral network involving public and/or private social service agencies should be established by an implementing organization whenever it observes that its clients could benefit from additional interventions. Finally, *collaborating organizations* are those whose duties include coordinating services very similar to those the new program plans to offer. For instance, a community-based organization working to implement a new antismoking program could benefit from working closely with a state health department, whose deeper pockets allow it to assist the program with training of implementers, technical updates, and additional resources.

5. Ecological Context: Seek Its Support

Like peer organizations/community partners, ecological context is a component of the action model with which only some programs need to be concerned. Ecological context is addressed in a program plan at the discretion of the working group or program designer. If a decision is made to include ecological context in the program theory or plan, both *micro-level* and *macro-level contextual support* may be important.

As to the micro level, working group members should ask whether the success of an intervention will depend significantly on support received from clients' social milieus or on adjustments to clients' physical surroundings. An education intervention, for example, will usually experience a better outcome when the parents of the targeted students are strongly supportive. Dependent on the nature of its program, a working group may need to consider the roles of clients' spouses/significant others, relatives, friends, neighbors, or coworkers. The group may also need to investigate target clients' physical surroundings: homes, schools, workplaces, and neighborhoods. The reason becomes clear in an example of an education program using innovative homework assignments to raise disadvantaged children's reading and math scores. Great benefit could accrue to this program if it provided each student with a home desk, because altering the physical surroundings this way should help foster routine habits of study. Or, think of the antigang program targeting juveniles in at-risk neighborhoods. The intervention message would be strongly underscored by certain changes to the physical surroundings: addition and/or repair of street lighting, restoration or removal of run-down structures, prompt erasure of gang-related "tagging" of fences and walls, and the overall sprucing up of sidewalks and streets.

Macro-level contextual support refers to the degree of potential support or opposition facing an intervention program from its community and its local institutions. The working group must certainly consider macro-level matters; if it detects lack of support (or active opposition) that could hinder implementation of the program, it should incorporate into the program plan some means of securing wider support for it. Some potential strategies for building macro-level support are media campaigns, consortiums, "summits" with opposition leaders, and community mobilization.

Media campaigns, including television or newspaper ads, draw attention to a problem and show people why a program is needed. Although media campaigns educate the public, they are unlikely to sway institutions or organization leaders who are publicly opposed to a program.

Joining or launching a *consortium* is a second support strategy. A consortium, or coalition of complementary organizations and agencies, exists to support and strengthen each other's agendas. Sometimes, the backing of a consortium helps win support for a program from the ecological context.

Summit-style meetings with leaders of the communities or institutions that oppose a program can also be support builders. In many situations, opposition to a program grows from misunderstanding of the program. It is sometimes possible, through exchange of ideas, to agree on a compromise that speeds along the implementation of the controversial program. The remaining strategy to be introduced here is *community mobilization*. It is grassroots and comprehensive, involving the recruitment and training of volunteers to contact systematically the public, opinion leaders, and officials of organizations and communities in an effort to generate support for a program. In a really well-mobilized community, former opponents can ultimately become involved in implementing the program.

6. Target Population: Identify, Recruit, Screen, and Serve

As Chapter 4 related, the program rationale identifies a target population or group, but the evaluator assisting a working group or program designer by employing conceptualization facilitation goes further, defining the target population absolutely and dictating strategies to recruit target population members to the program. Managing the target population component of the program plan can be an involved process. The reality of resource constraints on programs usually makes it impractical to target the entire population at risk of encountering a problem. Thus, although terms identifying a target population for a program rationale may be general ones, a precise specification or definition of the target population is ultimately needed. The program rationale of an HIV prevention program, for example, cites a target population of "individuals at high risk for HIV." There are, however, many groups of individuals at high risk of HIV exposure: immigrant workers, men who have sex with men, transgendered people, the homeless, sex workers, and intravenous drug users. Only a very wealthy program might attempt to serve them all. So, the program plan for this prevention effort must include clear eligibility criteria to be met by individuals before they are included in the target population. Often, these criteria reflect need or deficiency: low income, undereducated, abuse survivor, and so on. Alternately, eligibility criteria may rest on demographic characteristics such as age or residency. Program plans must be careful to avoid *overcoverage* or *undercoverage* of the pool of targeted individuals. Overcoverage results when eligibility criteria are too broad and loosely defined, allowing into the program those who do not really need its services.

An example of overcoverage is a government program intended to help small farmers that winds up serving significant numbers of well-to-do farmers with large farms, or even corporate farms, because its eligibility requirements were too elastic or imprecise. On the other hand, undercoverage denies eligibility to hosts of qualified members of a target population through the use of eligibility criteria that are too restrictive. Overcoverage can waste valuable resources, and undercoverage hamstrings the entire program. The evaluator can assist stakeholders in creating eligibility criteria that avoid both, achieving appropriate coverage.

The target population, once it is defined, learns about the benefits of a program through recruiting. Common strategies of recruiting include systematic marketing using radio, newspapers, magazines, pamphlets, and/or other mass media, referred to as a media campaign; referrals by agencies or organizations already serving members of the target population; and also outreach. An effective media campaign will hone in on places where target group members go about daily activities, including the neighborhood in which they reside. But for many programs, the target population alone is not enough. Often, outreach workers are ultimately needed to bring clients into programs. One popular outreach strategy engages former program clients as volunteer workers taking on "hardcore" cases or especially high-risk individuals. Former clients know where and how to reach potential new clients, and if they are volunteers, they can also conserve program resources; initially, though, the program needs to pay for at least some training of volunteers to ensure the quality of their outreach activities.

Target population members reached through media campaigns, referral, or outreach must almost always complete a diagnostic *assessment* first of all. Diagnostic assessment provides insight into an individual's unique problems, pointing the way to the kind of intervention and social services needed. The assessment is especially important before an individual is admitted to a more intensive type of intervention program (mental health care, substance abuse counseling/rehabilitation, etc.). Perhaps the major purpose of the assessment is to determine whether an intervention by the program will be beneficial for the client. Many health-related programs, especially, depend on careful diagnosis of each interested person to establish the medical necessity of the intervention. (Would-be clients discovered to not need an intervention after all, but likely to benefit from services beyond the assessing program's scope, can easily be referred to other agencies, in most cases.)

Once entered in an intervention program, clients are, unfortunately, likely to drop out again at some point, whether literally or figuratively. Excessive barriers to participation and low motivation are two reasons they do. A program plan that includes participant incentives begins to discourage this cycle. Two types of programs for which program designers need not worry about participation are mandatory programs and programs offering desirable goods or services to clients. Sometimes, interventions are available that are so obviously attractive that few clients can refuse. Programs providing benefits such as housing subsidies, food, medical care, or child care usually have no need to motivate clients. Whatever the target population is for such programs, its members likely will participate once informed that the program exists. Target populations for mandatory programs are also likely to participate because their participation is required by law. Mandatory programs include counseling for DUI offenders, traffic school for traffic law violators, and anger management classes for spouse abusers. Nonparticipation in such programs leads to financial penalties and even incarceration, so "no-shows" are few. But what of the free presentation on healthy eating and the drug use awareness program for teens that go virtually unattended? Intervention programs offer to do something good for a target population, but what is "good" is not always alluring. Incentives can help. When stakeholders and evaluators suspect that a treatment or intervention in and of itself will not secure the target group's participation, they should discuss using incentives. Fast food restaurant coupons have been used as an incentive to attend crime prevention programs, and drug prevention programs have offered tutoring and recreation along with their curricula. Every program needs to ensure that there will be incentive enough or motivation enough to engage the target group in the prescribed activities.

Programs need to remove barriers to participation whenever possible. Even a willing target population, reached and motivated, may be unable to fully participate in a program given certain barriers. Such potential barriers should be weighed by the working group as it prepares the program plan. Barriers can be those of *language, culture, stigmatization,* or *logistics.* If service providers do not speak and write the language of the target population, potential clients may be alienated. The cultural competence of program staff can also be an issue. Any target population has a racial and ethnic background, and sensitivity to that culture on the part of service providers can encourage retention of clients in a program. Another potential barrier to receiving some services is stigmatization of those services by the society. Participation in mental health services, for example, is viewed by many target groups as a sign of personal weakness or vulnerability. If this stigma is not overcome, target group members may decline mental health services despite implementers' best efforts. When stigmatization of services may be an issue, the program plan might suggest strategies for alleviating clients' fears; it might even outline some effort to change perceptions about the problem in question. A program plan may also need to address logistical barriers, such as lack of transportation or child care. Job training or drug treatment sites, for example, are probably not going to be conveniently located for all clients. Transportation needs are a particularly large factor in interventions with an indigent population. Although programs cannot, of course, buy cars for clients, where public transportation is available they can provide money for fares as one incentive to participate. If public transportation is not a good option, a program can provide a vehicle to pick up and drop off its clients. Program planners might even choose to "decentralize," providing services in the clients' own backyard. Like transportation, child care may need to be provided to indigent individuals if they are going to remain motivated to participate. The working group should discuss whether any of these barriers is likely to affect its target population.

Beyond barriers of a logistical or even cultural nature is the matter of indigent clients lacking basic needs such as shelter and food, and perhaps also suffering from physical or mental illness. Not infrequently, a client meeting all eligibility requirements may, in fact, be in no condition to receive a planned intervention. Program implementers find themselves providing referrals or even case management services to help such potential clients before (and occasionally during) the intervention. Larger programs sometimes dedicate a unit to provide these auxiliary services. For example, when a potential client of a sobriety program remains under the influence of alcohol, there is no way to deliver alcohol abuse treatment. The program that possesses its own detox unit may be successful in sobering up the individual, permitting participation in the program. The initial assessment of potential clients should make clear whether they need to become sober, and also whether they have a place to live, food to eat, relative good health, and adequate education. Although it is not the job of an intervention program to provide social services, a program plan frequently includes guidelines for linking eligible clients to relevant social service agencies. The person without a home and unsure if the next meal will be forthcoming is not likely to succeed in any intervention or treatment program. Thus, to be truly effective, programs whose target groups are susceptible to such material deficiencies must strive to connect these target populations with appropriate help.

APPLICATION OF THE CONCEPTUALIZATION FACILITATION STRATEGY

Example 1: A Garbage Reduction Program

In Taiwan, residents had been accustomed for four decades to placing garbage bags in designated pick-up areas each and every day. Daily pick-up resulted in huge amounts of garbage, to the detriment of the environment. A demonstration program proposed by the Neihou Sanitation Department was intended to determine if a new policy—no garbage collection on Tuesdays—would reduce the amount of household garbage set out for pick-up (Chen, Wang, & Lin, 1997). Program stakeholders engaged program evaluators to tell them, with holistic thoroughness, how effective such a policy would be. For the kind of outcome evaluation these stakeholders sought, an understanding of the program theory underlying the program was necessary. Because the program was newly begun, the evaluators had an opportunity to use the development facilitation strategy to assist in clarifying and developing the stakeholders' program rationale and program plan. To begin, the evaluators interviewed the program designers and other decision makers to learn about the program theory as understood by those individuals. Their basic notion of that theory, the evaluators found, hinged on the degree of unpleasantness they expected residents to experience when required to keep garbage in their homes even for one day (Taiwanese homes typically lack garbage disposals). Sufficient inconvenience and disgust, they believed, would raise awareness of excessive creation of garbage, resulting in new in-home efforts to reduce its volume. The program rationale and program plan that the stakeholders put together are illustrated in Figure 5.2.

Program Rationale: Target Population, Goal and Outcomes, Determinant, and Intervention

1. *Problem:* rapid increase in amount of garbage to be collected; ensuing environmental damage

2. *Target Group:* residents of the Neihou community

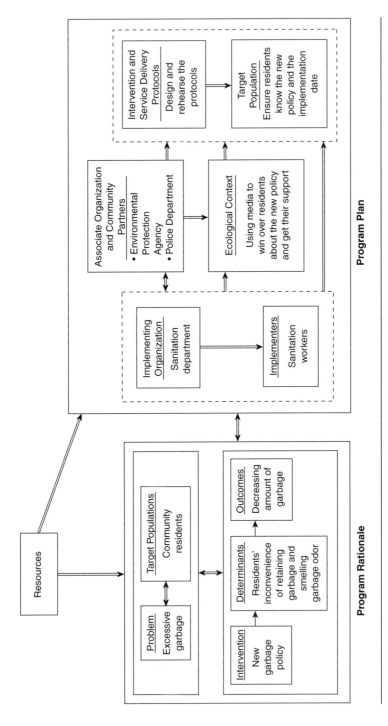

Figure 5.2 Program Rationale and Program Plan of a Garbage Reduction Program

SOURCE: Adapted from Chen et al. (1997).

3. *Goal:* decrease the amount of garbage to be collected

4. *Determinant:* residents' experiencing the inconvenience of retaining garbage in the home overnight and having to smell that garbage odor

5. *Intervention:* cease Tuesday collection of household garbage so garbage remains in homes overnight

Program Plan: Protocols, Implementers,
Implementing Organization, Peers and
Partners, Ecological Context, and Target Population

1. *Implementing Organization:* Neihou's sanitation department. The department's reputation was one of efficiency, and it had sufficient personnel and funding to implement the program.

2. *Program Implementers:* the department's regular sanitation workers, provided with training in implementation of the program

3. *Associate Organizations and Community Partners:* Taiwan's national environmental protection agency, the local health department, and the local police department

4. *Ecological Context:* media-based. To create awareness of and support for the new policy, stakeholders planned to advertise it in local newspapers and have the policy featured in a television interview.

5. *Intervention and Service Delivery Protocols:* prohibition by law of disposal of household garbage on Tuesdays, to be enforced with fines. To prevent dumping at garbage collection points on Tuesdays, sanitation department employees would patrol these sites on Tuesdays, empowered to warn and then cite violators, who would be fined for noncompliance. Pilot testing of the protocol was to take place before implementation.

6. *Target Population:* residents of the Neihou community. All affected residents were to be notified of the new policy via two waves of informational letters sent to them by the sanitation department. In addition, huge banners reminding residents of the policy were to be displayed on all major roads in the area, both prior to and during its implementation.

Advantages of Development Facilitation as
Illustrated in This Example

The stakeholders expressed that the development facilitation approach was useful in assisting them with the conceptualization of their program. They felt that the approach systematically connected their thoughts about the program. The program rationale and program plan had been used by stakeholders to communicate the program to audiences inside and outside the community. The program plan was particularly helpful to them in ensuring the quality of implementation. The program rationale and program plan were used as a foundation for evaluators to design and conduct theory-driven process and outcome evaluations (Chen et al., 1997).

Example 2: An HIV Prevention Program

A funding agency concerned about the spread of HIV launched a program planning process by offering grants to community-based organizations (CBOs) willing to undertake new HIV prevention efforts in the African-American community. This agency brought in program evaluators, who suggested using the conceptualization facilitation strategy to help the agency's stakeholders conceptualize their grant program (Chen, 2003). The funding agency required that each participating CBO incorporate at least the following four program activities in its program plan:

1. Through means consistent with applicable existing state and local HIV prevention plans, provide high-risk individuals with (or assist them in gaining access to) HIV counseling and testing, as well as with appropriate referrals for other needed services. A proposed program not strictly consistent with the more comprehensive prevention plans must adequately justify its intention to address other priorities.

2. Conduct primary HIV prevention activities such as health education and risk reduction interventions for people at high risk of becoming infected with HIV or transmitting the virus to others.

3. Assist HIV-positive people in gaining access to appropriate early medical treatments, substance abuse services, STD screening and treatment, perinatal health care, partner counseling and referrals,

psychosocial support services, mental health services, tuberculosis prevention/treatment, and other supportive services as needed. Refer high-risk clients who do not test positive for HIV for meaningful health education and risk reduction programs and/or other appropriate prevention services.

4. Frame all program activities and intervention messages with sufficient cultural competence and linguistic and developmental appropriateness.

The evaluators were asked to consider these requirements and facilitate a working group's planning of the grant program implementation at multiple sites. The working group included a manager and three staffers charged with bringing the grant program to fruition, plus the evaluators. The group's task was to develop a coherent program plan from the brief terms of the grant announcement. To facilitate discussion, before the working group met, the evaluators compiled a rough draft of a program plan based on the four requirements and limited informal talks with the stakeholders. At its first meeting, the working group developed a program rationale and program plan, both of which follow.

Program Rationale: Target Population,
Goal and Outcomes, Determinant, and Intervention

1. *Problem:* the upward trend in HIV transmission among minority groups

2. *Target Population:* African-Americans at high risk of HIV exposure or already HIV positive

3. *Goal:* slow down HIV transmission in the target group, which has been especially susceptible to the HIV epidemic

4. *Determinant:* better access to HIV prevention and treatment services for African-American clients who, like others in that minority, lack avenues to existing HIV testing and counseling outlets

5. *Intervention:* provide money, technical assistance, and capacity-building services to CBOs working to reach African-Americans and guide them to HIV prevention and treatment services

Program Plan: Target Population,
Protocols, Implementers, Implementing
Organization, Peers and Partners, and Ecological Context

1. *Implementing Organization:* a CBO willing to accept capacity-building and technical assistance at the discretion of, and paid for by, the grant agency

2. *Program Implementers:* people demonstrating both technical and cultural competence

3. *Associate Organizations and Community Partners:* entities involved in state and local health planning affecting the CBO, plus other relevant bodies (national, regional, state, and local) supportive of HIV prevention/treatment efforts and capable of facilitating the CBO's client referrals or curtailing duplication of efforts

4. *Ecological Context:* CBO needs to conduct needs assessment and seek community inputs and support of its services

5. *Intervention and Service Delivery Protocols:* to provide CBOs with protocols in counseling and testing, and referral

6. *Target Population:* HIV positive and high-risk African-Americans, prioritized according to local trends (determined by needs assessment, epidemiological profiles, and state/local anti-HIV plans): intravenous drug users, men who have sex with men, sex workers, the homeless, and so forth, with *each* risk group to be reached via a tailored, concrete, CBO-designed targeting strategy

The working group decided to place the strongest emphasis on those portions of the program plan concerned with reaching the populations within the target group and linking its members to needed services (see Figure 5.3).

The working group briefly discussed the type of outcome evaluation the CBOs should conduct following implementation. It reached a consensus, maintaining that such an evaluative task was beyond the CBOs' current capacities. Instead, CBOs would be assigned simply to collect certain evaluation data for their own use and to forward to the funding agency, primarily to meet accountability needs. The working group settled on the following list of probing questions:

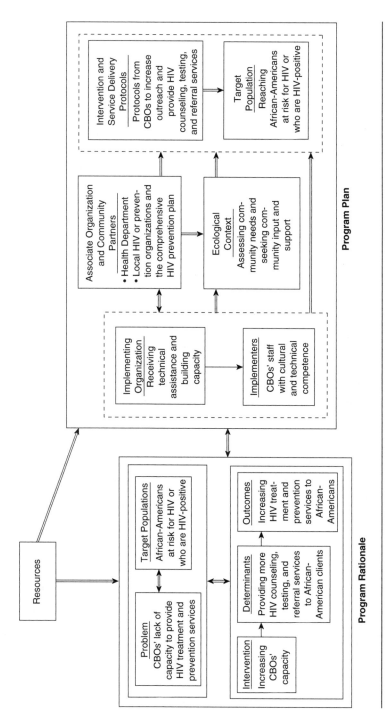

Figure 5.3 Program Rationale and Program Plan of an HIV Prevention Program

SOURCE: Adapted from Chen (2003).

- Does the program reach African-Americans in high-risk populations?
- Does the program reach the target population?
- How many target population members have been served by the program?
- What kind of services have they been provided?
- What has been the quality of these services?
- Have clients referred for treatment actually received treatment?
- How many clients express satisfaction with the program?

Advantages of Conceptualization
Facilitation as Illustrated in This Example

Conceptualization facilitation as deployed in the preceding sample of program planning presented five main advantages to stakeholders. First, it produced for them a coherent, systematic program rationale and coherent, systematic program plan that were consistent with each other. It fostered stakeholders' understanding of, and ability to debate, the essence of the announced requirements of the grant; this helped ensure that the requirements would be reflected in the program plan. Second, the conceptualization facilitation process revealed assumptions implicit in the program rationale that needed to be scrutinized (and, where necessary, turned this scrutiny into action to strengthen the program plan). Looking over the program rationale and plan (reproduced here as Figure 5.3) prompted the working group to discuss service routes. The discussion focused on CBOs and their prevention workers, who had been placing too much emphasis with outreach clients on prevention services activities and not enough emphasis on HIV counseling and testing. Only when the high-risk client's HIV status becomes known, through testing, can the appropriate services be determined. Strategies to increase stress counseling and testing by CBOs were discussed.

A third advantage of the facilitation process was its usefulness in linking the new CBO-managed activities to related activities alluded to in other funding announcements. For example, separate announcements were issued concerning capacity building and coalition development for CBOs. The model (Figure 5.3) indicates graphically a clear need to integrate the various announcements. A fourth advantage of the facilitation process was the opportunity it provided for programming staff and evaluation staff to firmly understand the kind of evaluative information needed at given program stages. This understanding was a foundation from which to develop evaluation indicators—and an excellent evaluation design. The fifth advantage of the conceptualization facilitation strategy used here was a by-product of the process: insights as to promising foci for future programs. For example, although the grant announcement cited in this example was silent on the issues of removing barriers, motivating clients, and building contextual support for intervention, the model of conceptualization facilitation shows that these are, indeed, important domains deserving incorporation within program plans. Thus, the working group's discussion of future program planning was a conversation that has likely informed each member's work with program plans.

THE PILOT-TESTING APPROACH

The pilot-testing approach falls under the troubleshooting strategy in the taxonomy of program evaluation means and ends (see Table 3.1). Pilot testing determines the field feasibility of a program plan, and thus is a way to avert the implementation of a significantly flawed plan. Often, once a program plan is completed, the program director and implementers feel pressure to implement the plan immediately. Pilot testing is valuable when these stakeholders want quick feedback to use in fine-tuning their plan before implementation. Because timeliness is likely to be paramount, a pilot test must be flexible as to research methods; this flexibility even extends to testing only parts of a program rather than the entire program. More evidence of pilot testing's flexibility is its use of samples much smaller than those demanded by traditional evaluation. Strictly speaking, the rapid inquiry of the pilot test is more a development tool than a formal assessment tool.

Defining Pilot Testing

The term *pilot testing* is often associated with tests of the reliability and validity of questionnaires or measurements. In this book, however, the meaning of pilot testing is not measurement alone. Rather, it refers to a feasibility study, a small-scale field trial of a program plan conducted rapidly—over a few weeks or months—in order to assess and improve the implementation of the full-scale intervention. Part of the rapidity of the approach is the timely analysis of data and presentation of findings to stakeholders by the evaluator. A pilot test can include testing of instruments, but its main purpose is discovering problems that might arise during implementation of a program plan. The characteristically small sample size involved in pilot testing can shrink to just a handful of subjects when the program under review focuses on hard-to-reach clients such as homeless people and drug addicts.

Furthermore, it is important to distinguish pilot testing from related concepts like *pilot study* or *demonstration project.* More time and more topics are involved in these than in pilot testing. Pilot studies and demonstration projects characteristically test all aspects of a program and also cross the various stages of that program, extending to testing the program's effectiveness. Their ultimate concern with program effectiveness encourages the use of quite rigorous methods by pilot studies and demonstration projects and, in turn, the diversion of quite significant resources.

Conducting Pilot Testing

Four principles are especially significant to the program evaluator preparing a pilot test to generate information that can improve program plans. Those principles are as follows:

1. *Pilot testing requires actual implementers and clients to participate in trials.* Feedback from the very implementers involved deeply in the day-to-day activities of the program offers firsthand information about looming problems with implementation, as well as educated guesses about managing problems. Feedback from the clients whose very lives can be changed by a program produces genuine insight into what makes, for them, a satisfying intervention delivery. In earlier chapters, it was emphasized that program plan development is the realm of experts and top officials of organizations; however, actual implementers and clients may have their own, very different and very valuable, perspectives about a program.

2. *Pilot testing relies on small but nevertheless typical samples.* To keep the feedback prompt, pilot testing usually relies on a small group of clients and implementers. These participants should be typical of their groups for best results. For example, if a majority of clients for a job program are expected to be drawn from the persistently unemployed, then the handful of clients in the pilot test should also be persistently unemployed. If, instead, these pilot-test subjects were relatively experienced in keeping a job, the pilot-test results would be highly misleading. A good rule of thumb for pilot testing is to avoid extremes, such as overly enthusiastic or apathetic target group members, or overachievers or underachievers among their ranks. Another rule of thumb is

that when a sample implementing organization is needed for the pilot test, select an organization typical of the ones that will conduct the full implementation. If mom-and-pop CBOs will implement the program, do the pilot testing with a mom-and-pop organization, because results obtained for a large, complex, sophisticated organization would be difficult to generalize to the complete program in its planned form. The same principle should be applied when pilot testing a delivery mode, ecological context, or other component.

3. *Methods of gathering data in a pilot test must be flexible.* With a small sample and short time frame, the pilot test is, in most instances, better off collecting its data with methods like the focus group, interview, survey, site visit, and so on. There is no need to apply rigorous methods like the randomized experiments typifying the pilot study or demonstration project.

4. *Pilot-test findings should be used only for program development purposes.* Pilot testing is of a developmental nature. Because it tests only parts of a program, pilot-testing results should never be used as evidence of a program's effectiveness. The pilot test asks whether a program shows signs of major implementation problems, not whether it will be effective in the end. Interpreting the results of pilot testing is made simpler by this rule of thumb: When a pilot test suggests a program plan is substantially flawed, full-scale implementation of the plan likely will also be marred by substantial flaws unless they are resolved in the program plan beforehand. On the other hand, when pilot testing suggests the program plan works well, the plan may—or may not—generate a successful full-scale implementation. Positive results from pilot tests simply are not indicators or evidence that an implementation will be of high quality or that an intervention will be effective. Such evidence comes from process evaluation and outcome evaluation, not pilot testing.

Designing Pilot Testing

The key to designing pilot tests is to mimic exactly the program activities and processes planned for the full-scale implementation. For example, for interventions that, when implemented, will comprise several sessions over a period of time, the pilot test ideally would involve an identical schedule: same session length, same day of the week, same time span, and same setting. The action model framework can be consulted as a systematic means of considering each component of the program plan to be tested. It is up to the stakeholders to determine just what information will be collected via a pilot test.

The evaluator may present them with certain guidelines, however, for selecting or passing over topics.

In general, the following questions need particular attention during pilot testing:

- Can the intervention be implemented in the field as intended?
- Can implementers anticipate encountering certain problems delivering the intervention?
- Will clients be receptive to the intervention or resist it?
- Do any of the program's organizational procedures impede the implementation process?

Intervention and Service Delivery Protocols. Protocols are the components most frequently evaluated by pilot testing. They are tested by actually delivering an intervention to clients in the planned setting. Although comments from implementers and clients are something of a reality check, evaluators can also choose to observe in person the delivery of a service, watching for potential problems. When clients and implementers are queried in the course of pilot testing, the following issues should be brought out in interviews or surveys:

- Is the dosage of the intervention satisfactory?
- Does the intervention require too much time? Conversely, is it too brief to be effective?
- How much difficulty do implementers experience in delivering the service?
- Do clients find it difficult to follow the language or procedures of the intervention?
- Does the intervention setting help or hinder service delivery?
- Is the intervention schedule amenable to clients and implementers alike? To illustrate, consider a day care center for working mothers, sponsored by a welfare program that has chosen to close the center at 5:30 p.m. each day. Because many of the women use public transportation and deal with its vagaries, this schedule is not amenable to them.

Deciding how to implement an intervention in the field is one of the most challenging tasks facing a new program. Prospective clients and implementers alike can contribute to the decision by giving feedback as to the practicality of the planned mode of service delivery for the planned intervention in the planned setting. For example, one program plan called for hiring a professional outreach worker to bring a prevention message to a high-risk neighborhood. But when speaking with a handful of prospective clients during a pilot test, the program evaluator heard their opinion that an outsider would be extended little opportunity to interact with the neighborhood's residents. This finding prompted revision of the program plan, which eventually stipulated the involvement of outreach volunteers from the neighborhood.

Target Populations. The reaching and recruiting of members of a target population is another important focus for pilot testing. No program will survive for long without clients, and a pilot test is an opportunity to ensure that clients can be obtained via a proposed recruitment strategy. Take as an example a child abuse prevention intervention targeted at parents living in inner-city high rises. Its proposed recruitment strategy is complicated: the cooperation of building managers is required, and doors must be knocked on one by one to introduce households to the program and extend individual invitations to participate. Stakeholders who decide to pilot test the recruitment strategy can learn whether building managers support the program, whether parents are home during the day, whether the population of young children in the buildings is significant, and whether parents feel open to the notion of attending the program or would feel stigmatized by it. If pilot testing shows the strategy is likely to fail, stakeholders have a chance to devise a stronger strategy. This alone could be the difference between a foundering program and a thriving one following full-scale implementation.

When program planners see a need to use incentives to recruit clients, a pilot test can show whether the selected incentives will actually motivate individuals to complete the program. Is a storybook incentive enough to bring children to dental health sessions? Pilot testing should be able to answer the question quickly. (Sometimes, the results can be unexpected.) Any barriers to program participation can also be scoped out through pilot testing. A mental health program for the poor, for example, could discover through pilot testing that eligible people believe that there is a stigma with such services, discouraging them from participating, which is something unaddressed by the program plan.

Implementing Organization and Implementers. Pilot testing an implementing organization is more difficult than testing other action model components.

In many cases, evaluators resort to site visits instead. Such visits are an opportunity to gather data about organizational capabilities, *indirectly* approaching the issue. General inquiries during a site visit could include the following:

- Does the implementing organization have the skills, resources, and commitment necessary to implement the intervention?
- Are the implementers qualified to deliver services?
- Can the implementing organization build and sustain collaborations or linkages with related organizations to facilitate delivery of the intervention?

Ecological Context. Sometimes, client participation hinges on the ecological context of a program, notably whether family and friends support an individual's enrollment in that program. Stakeholders must ask if such support can be counted on, and a pilot test can often give an answer. For instance, in a plan for a delinquency intervention program for juveniles, if it is acknowledged that parents' involvement and support are key to the success of the program, pilot testing this ecological context makes a good safeguard. The pilot test will indicate whether parents of the targeted juveniles are supportive. After analyzing and compiling the data generated by observation and interviews, the evaluator works with stakeholders to revise and finalize the program plan prior to implementation.

A good example of the course of pilot testing is found in a school-based cancer prevention program, "Key to Health," which proposed using five weekly, 90-minute sessions in which teachers taught adolescents about the role of a low-fat, high-fiber diet in preventing some cancers, and supported their efforts to follow such a diet (Wallin, Bremberg, Haglund, & Holm, 1993). A pilot test saw two portions of the Key to Health curriculum offered to students during a 3-month period. Essentially, these test sessions gave the students an opportunity to become aware of and reflect on their own eating habits and perhaps experience a healthful, self-initiated change in diet. Each student set an individual dietary goal and made changes based upon it. At the end of the third month, information about the program's operation was collected from those teachers and students who had joined in the test sessions. During this pilot test, 49 students completed self-administered questionnaires, eight students joined a focus group moderated by a program evaluator, and the teachers completed both structured interviews and a survey. Results suggested that the tested Key to Health program had worked well and could be integrated quite well into existing curricula (in a home economics course, perhaps). But the pilot testing also indicated a discrepancy between students' and teachers' views of self-efficacy as a force behind healthful dietary changes. It was decided by Key to Health stakeholders that any setbacks likely to stem from this discrepancy could be overcome with teacher training.

THE COMMENTARY OR ADVISORY APPROACH

Evaluators are not always solicited to provide formative research or development facilitation; they (or other experts) may be asked by stakeholders simply to comment on a completed program rationale and program plan. The stakeholders are especially likely to seek advice for improving these. In a situation like this, program evaluation occupies the role of troubleshooter; indeed, the commentary approach falls under the troubleshooting strategy in the practical taxonomy (see Table 3.1). The conceptual framework of program theory, including the change model and action model alike, is a beneficial tool for evaluators when it comes to commenting and advising. Looking to it, a number of questions come to mind whose answers are clues to the quality of the program rationale and plan. Evaluators might ask this set of questions to locate those points on which they should comment.

Questions Advancing the Evaluator's
Commentary on a Program Rationale

- Does the program rationale specify target populations and systematically explain the intervention, determinants, and goals/outcomes?
- Are all elements present in the rationale adequately specific and apparently justifiable?
- Do the relationships that the rationale assumes to exist among these elements stand up to scrutiny?
- What procedures have been used to ensure various stakeholders' understanding and support of this program rationale?

The program rationale itself might be the first question an evaluator asks when critiquing a program plan: Does the program rationale exist to guide the development of the program plan? If not, the evaluator should steer the stakeholders to adopt a program rationale that can be drawn upon to revise their program plan. The following recommended *further* questions are generated from the action model conceptual framework.

Questions Advancing the Evaluator's Commentary on a Program Plan

About the Implementing Organization

- Does the implementing organization have the experience and the capacity to implement the intervention?
- Is the implementing organization experienced in working with the target group, and is it sensitive to this group's culture and needs?
- Has the implementing organization earmarked the necessary resources and personnel to implement the intervention?

About Intervention and Implementation Protocols

- Does the program plan include an intervention protocol specifying which curricula and activities the intervention will comprise?
- Does the program plan include an implementation protocol specifying the setting for service delivery and the procedures to use in delivering services to clients?

About Implementers

- Have reasonable minimum qualifications for implementers been clearly specified?
- Will training be provided for implementers?
- Have procedures been planned to ensure implementers' cultural competence (e.g., training)?

About Associate Organizations/Community Partners

- Has the implementing organization identified pertinent organizations with which it might profitably collaborate on service delivery?
- Has the implementing organization developed strategies for working with associate organizations to facilitate service delivery?

About Ecological Context

- Will support from family, friends, and/or coworkers be required for clients to succeed in an intervention?
- If such support is required, what strategies and procedures are in place to secure it?
- Will community support of the intervention program be required for the program to work?
- If it will, are strategies and procedures in place to secure such support?

About Target Populations

- Are the eligibility criteria for potential clients clearly defined and practical to implement?
- Have precise and feasible strategies been developed to reach the target populations?
- Are sufficient incentives in place to persuade clients to participate in the program?
- Does the program adequately understand and address the barriers that may come between target population members and program participation?
- Can the program recruit, with reasonable effort, a sufficient number of clients?

SUMMARY

The four evaluation approaches that have been covered in this chapter are means by which an evaluator can assist stakeholders as they develop program plans: formative research, conceptual facilitation, pilot testing, and commentary or advisory. Formative research generates background information that stakeholders can use in creating a program plan. Conceptual facilitation allows the evaluator to facilitate the stakeholders' task of clarifying or developing a program plan.

With a program plan already in hand, the approaches most useful to typical stakeholders are pilot testing and commentary or advisory, both of which work to fine-tune a program plan. The commentary approach is the low-cost alternative but has a disadvantage in its lack of hard data from the field. Even experts' comments and suggestions can be too general and subjective. Although useful in its own way, the commentary approach does not work as a substitute for pilot testing when pilot testing is needed. The most important advice this book can give along these lines is to use pilot testing—if *at all* possible— to fine-tune any start-up program.

A Place for Process/Outcome Evaluation in the Program Plan

When program evaluation is part of a program from its earliest planning stage, it is a major advantage. Evaluators bring broad expertise to the table; it is know-how that extends even to stakeholders' distant evaluation needs, those arising in the final stages of program development and after. Process evaluation and outcome evaluation are the options most likely to meet these late-stage needs, so in some circumstances it may make sense to include in the program plan a schedule of future processes and/or outcome evaluation (perhaps along with a statement of stakeholders' expectations for such evaluation). The details of process evaluation and outcome evaluation design are discussed elsewhere in the book.

Another Strategy From the Practical Taxonomy

The taxonomy of practical program evaluation features one evaluation strategy appropriate for planning-stage evaluation that, up to this point in the book, has not been discussed at length. It is called *development partner strategy.* Sometimes, a group of stakeholders, realizing its program is quite fluid and undefined, and desirous of a great deal of input, may invite a program evaluator to become an actual partner in the planning and development process. In action, the development partner strategy frequently operates from the initial planning stage through implementation. Guidelines for applying this strategy are presented in the next chapter.

⊰ PART III ⊱

EVALUATING IMPLEMENTATION

—•—

Implementation is where the action is in programs. An implementation must be a success before a program can be considered effective; without appropriate implementation of the program plan, there can be little expectation that the program will succeed. Implementing a program plan is usually a complicated and challenging course of action. Program evaluation makes a good tool for stakeholders who want to ensure that the implementation of their program is being carried out as intended. Evaluation of implementation processes has been called implementation evaluation or process evaluation. This book uses implementation evaluation as a general category for covering both assessment- and development-oriented evaluation of implementation process and reserves the term *process evaluation* for those implementation evaluations carried out at the mature program stage, which will be discussed in Chapter 7. In discussing implementation evaluation, it is important to note that implementation has two stages: the initial implementation and the mature implementation. Distinguishing the stages from one another is vital because stakeholders' evaluation needs are distinct at each stage. What stakeholders are seeking at the initial implementation stage is quick feedback for use in developing their program; thus, the evaluation information obtained is largely for internal purposes. In evaluating the initial implementation, strategies, approaches, and research methods need to be flexible and expeditious. Development-oriented evaluations are useful in this stage.

In contrast, in the mature implementation stage, evaluative information is needed to meet accountability requirements (as well as to shape stakeholders' discussion of long-term strategic improvement of their program). External audiences such as funding agencies expect stakeholders to provide credible evidence of the success of a program's implementation. For this reason, evaluation procedures and research methods used to assess mature implementation must emphasize rigor. Scientific rigor typically requires a relatively greater investment of time, so evaluations of the mature implementation stage usually take longer to design and conduct. When stakeholders approach an evaluator about process evaluation, the parties initially need to discuss the stage of implementation in question and the purpose for the process evaluation. Communication about these two aspects, in particular, will allow the selection of an appropriate evaluation strategy and approach for the task. In Chapter 6, the strategies and approaches appropriate for evaluation of an initial implementation are presented, whereas Chapter 7 focuses on procedures that are best when evaluating a mature implementation.

SIX

DEVELOPMENT-ORIENTED EVALUATION TAILORED FOR THE INITIAL IMPLEMENTATION

A n evaluator brought in to evaluate a program during the initial planning stage may have previously evaluated the program's plan. If so, his or her familiarity with the program is a boon for the design of the latter evaluation. If not, it is well worth the investment of the evaluator's time, prior to designing and conducting the evaluation, to learn how stakeholders have conceptualized their program.

The initial implementation stage remains a formative one in which the program is fluid. The procedures and rules governing implementation are frequently revised. During this stage, many programmatic concerns are being addressed simultaneously: recruiting and training implementers, establishing service delivery procedures, reaching potential clients, contacting related organizations, dealing with unexpected crises, and so on. We have seen how a high-quality program plan and the use of pilot testing reduce the number of problems encountered. Even with these safeguards, however, difficulties can arise. When having the initial implementation evaluated, stakeholders are looking for quick feedback about significant setbacks with implementation, in the hope of overcoming them efficiently and precluding damage to the program as a whole. The type of evaluation they want is *development-oriented evaluation.* In the program evaluation taxonomy (Table 3.1), development-oriented evaluation corresponds to two strategies that are highly recommended for evaluation of initial implementations: the *troubleshooting strategy* and the *development partnership strategy*. The troubleshooting strategy returns feedback from the field rapidly in order to report suspected problems with the implementation and give insight into possible solutions. The development partnership strategy is a response to stakeholders' concerns that a program and implementation may be overly fluid and vague; it involves a program evaluator functioning as a partner in development work.

As indicated in the taxonomy (Table 3.1), the troubleshooting strategy covers two evaluation approaches: the *formative evaluation* and the *program review/development meeting*. A formative evaluation can be used by the evaluator to collect firsthand information about potential implementation problems and their management. The program review/development meeting is more useful for building consensus among program staff as to the implementation problems encountered and the remedial actions available. In slight contrast to the troubleshooting strategy, the development partnership strategy and approach fit any program that seeks frequent input from evaluators during its planning and development phases. Below, the nature of each strategy and approach, and its pros and cons, are discussed. The discussion provides a guide to selecting and applying appropriate procedures for evaluation of initial implementation.

THE FORMATIVE EVALUATION APPROACH, UNDER THE TROUBLESHOOTING STRATEGY

Troubleshooting is a rapid-fire strategy for developing stakeholders' awareness of major problems in, or barriers to, the implementation of a program. The troubleshooting strategy allows evaluators to systematically and rapidly gather information about the existence and possible sources of problems; it also equips them to facilitate stakeholders' efforts to remedy problems. Troubleshooting can be conducted at any point following a program's implementation, but is especially useful during the initial implementation stage. One popular evaluation approach associated with this strategy is *formative evaluation*. Formative evaluation employs flexible research methods to assess barriers to and facilitators of implementation, enabling stakeholders to troubleshoot problems. Formative evaluation is different from the *formative research* discussed in Chapters 4 and 5. Formative research provides background information to further stakeholders' design of a program, whereas formative evaluation is a development-oriented evaluation applicable once a program is formally implemented. Because formative evaluation is conducted only with formally implemented programs, it is also different from pilot testing as described in the preceding chapter. In fact, large and complicated intervention programs usually need formative evaluation *and* pilot testing. (Generally, small and/or straightforward programs require only one or the other.) Pilot testing involves pieces of programs rather than whole ones, or it tests smaller-scale versions of the proposed program. Although pilot testing is useful in preparing for initial implementation, formative evaluation is still needed in this stage to ensure the success of the implementation.

Timeliness and Relevancy

Good formative evaluation must meet two criteria simultaneously: timeliness and relevancy. Timeliness means that an evaluation keeps to the stakeholders' time frame. Because stakeholders are, at this point, in need of quickly collected information they can use to continue efficiently down the road toward mature implementation, an evaluation that is too slow paced will not serve. A great deal of the value of formative evaluation lies in its ability to present information to stakeholders quickly. The value of formative evaluation also resides in its ability to identify crucial implementation problems likely to affect the quality of the program overall. The action model conceptual framework shows evaluators where to focus in their search for such problems.

Research Methods

To ensure its relevancy and timeliness, formative evaluation tends to employ research methods that are flexible and can be tailored to particular evaluation circumstances. Research methods popular for formative evaluation are focus groups, participant observation, key informant interviews, and small-scale surveys. These suit the necessary focus on programmatic inquiries: Can implementers reach the intended clients? Are implementers having difficulty delivering services? Are clients receptive to the intervention? Does the community support or oppose the implementation? The answers to these questions direct stakeholders in fine-tuning and managing the program for success down the line. The results of formative evaluation are frequently adopted by program directors and implementers to revise the implementation process in the initial implementation stage. During initial implementation, program structures are not yet firmly established, and modifications to the program are relatively easy to make. Formative evaluation provides timely information to serve programmatic needs.

A hypothetical program demonstrates the usefulness of timely, relevant formative evaluation. The clients of a fairly new family counseling program for Asian immigrants are frustrated with what they perceive to be counselors' lack of understanding of Asian cultural beliefs and family values. Evaluators can document this problem quickly through a formative evaluation that uses either focus group meetings or interviews of the immigrant clients. Detected quickly, the problem can also perhaps be solved quickly. Trying to use time-consuming, large-scale, rigorous research methods with a large representative sample could result in an evaluation that finally wraps up *after* numerous clients had become so frustrated that they opted out of the program. There is a place in process evaluation for rigorous design and research, of course, but it is not in the initial implementation stage. (Chapter 7 will explore the comparative rigor of process evaluation during the mature implementation stage.)

Steps in Applying Formative Evaluation

The formative evaluation approach is applied in six basic steps. They are presented here, along with general principles of formative evaluation pertaining to each one.

1. Review Program Documents and Note Underlying Assumptions

To be sensitive to issues involved in the development of a particular program, evaluators must know the program, and its purpose, in detail. Evaluators who participated in making a program rationale and program plan have an advantage when it comes to carrying out formative evaluation of that program: They are aware of the assumptions that underlie the stakeholders' decisions. However, the evaluator invited to carry out formative evaluation after the closing of the planning process can acquire the necessary information from documents and interviews with stakeholders. Mutual understanding between evaluator and stakeholders is the objective of the interviews and document analysis. Without mutual understanding, the quality of the formative evaluation is in jeopardy.

2. Identify the Program Elements Crucial to Successful Implementation and Determine Which May Be Vulnerable

The need for timeliness makes it impossible, during process evaluation, to examine every aspect of a program. Evaluators need to work with stakeholders to determine which parts of the program are likely to be most vulnerable and deserve additional attention. Using the action model conceptual framework, the evaluator can facilitate brainstorming by stakeholders about vulnerabilities that may call for an intensive check.

3. Select Well-Suited Data-Collection Methods

Again, formative evaluation uses research methods that are flexible and provide quick feedback. Participant observation, key informant interviews, focus groups, site visits, record reviews, and small-scale surveys are some popular tools. Often, an evaluator must tailor the research method somewhat to fit the evaluation circumstances. For example, when surveys are employed in formative evaluation, they may not embody ideal survey methodology—notably, the use of a large, representative sample. To work within the stakeholders' time frame, smaller samples must do. Formative evaluation deals with programmatic issues, which are relatively robust and easily documented (unlike proof of causal links between variables, the object of traditional research). Within the field of program evaluation, information from focus groups and small survey samples is perfectly capable of raising the necessary red flags if real problems are facing a program. For example, a focus group voices a majority opinion that waiting times for services are too long and frustrating, and adds that implementers' rudeness is fueling resentment. Whatever methodological limitations "taint" the information, it still points to a problem needing immediate attention.

4. Identify Problems

In the information the evaluator obtains, any elements or activities of the implementation suffering difficulties should show themselves. The findings can quickly be related to the stakeholders.

5. Probe for Sources of Problems to
Help Stakeholders Choose Remedial Action

Formative evaluation becomes more useful to stakeholders when it goes on to provide information about a problem's source and strategies that might resolve the problem. This is why probing the reasons for a problem is an important part of a formative evaluation. For instance, the evaluator who finds clients unreceptive to implementers will immediately ask why they are unreceptive. Do they feel ill at ease with implementers because implementers seem inadequately trained, or insensitive to culture and language, or overly hurried? Answers to the evaluator's probing questions generate timely feedback, providing a base for remedial action by stakeholders.

6. Submit Findings to Stakeholders and
Document Changes They Make Based on the Findings

It is likely that results of formative evaluation will prompt stakeholders to modify the program; ideally, formative evaluation does what is needed to ensure the program is implemented appropriately. The emphasis on action contributes to a tendency among program staff to be less than diligent about documenting modifications approved for the program plan. Eventually, the written program plan and other documents no longer reflect the reality of stakeholders' intentions. Such neglect can create many problems when a program is later assessed on the merits of its implementation or on its effectiveness. It is highly desirable for evaluators to work with stakeholders to immediately and systematically document all important changes and revise the program plan.

Four Types of Formative Evaluation

Formative evaluation frequently takes one of four forms: on-site observation and checking, focus group meeting, intensive interview, and systematic scanning. Below, the nature of these is discussed, and steps involved in employing them are listed.

1. On-Site Observation and Checking

On-site observation and checking involves the evaluators themselves participating in a program, or else observing the implementation process, in order to identify major implementation problems (if any) and probe their sources. This kind of observation entails witnessing the service delivery processes and then also interviewing clients and implementers. Program directors and staff are quickly informed of findings to facilitate their decision making. On-site observation and checking cannot be done without some preparation. The evaluator does not simply jump into the field with no conceptual grasp of the program's intentions and limits just to observe whatever transpires. To get useful information to stakeholders, evaluators need to be familiar with stakeholders' ideas about the program plan. The action model conceptual framework can help the evaluator systematically review these ideas with the stakeholders. In the course of discussion, the evaluators should probe stakeholders about areas of the program they might consider to be weak and subject to potential implementation problems. This is an important task because it ensures that, at a minimum, the evaluation will provide the information of most interest to stakeholders. However, the evaluator should not feel confined to only those potential weaknesses remarked on by stakeholders. Once in the field, the evaluator must feel free to investigate other issues emerging from observation or interview.

During on-site observation and checking, there is no need to complete all checks of potential problems before communicating with stakeholders. Time is of the essence, so when an implementation problem (and perhaps its likely source) has been identified, the evaluator passes the information along immediately to the program director and implementers. To withhold the information would be to deny stakeholders time they could have used to develop a resolution. Nor is there any requirement that on-site observation and checking be a one-shot evaluation. The approach can be applied for as long as needed; in general, as compared to smaller programs, larger programs take longer to reach a state of mature implementation, meaning on-site observation and checking of a larger program will cover a longer period.

An Example of On-Site Observation and Checking. Shapiro, Secor, and Butchart (1983) studied a leadership and management training program designed for women working in higher education; their effort provides a good example of useful on-site observation and checking employed to strengthen a program in its initial implementation stage. The training program's purpose was to develop positive self-concepts in the women that might help move them into administrative positions. Most participants were support staff, entry- or mid-level administrators, and faculty. The studied interventions included brown-bag seminars, a case study workshop, and a leadership and management clinic, which evaluators joined as participant-observers. Evaluators also conducted surveys of clients' satisfaction with these activities. The evaluators identified three major problems and described them for the program developer in ad hoc summaries. The problems were segregation within the target group, confusion about co-facilitators' duties, and a feeling that presentations were somewhat nebulous. At the first meeting, participants sat with and worked with women in the same occupation; most viewed this as elitism. Made aware of this difficulty, program staff used mixed groups in subsequent meetings. The sense of segregation and elitism was no longer an issue after this change.

Furthermore, program stakeholders had hoped that the co-facilitators they hired would evolve into strong leaders. But in interviews with the co-facilitators, evaluators found that they felt unprepared for their tasks because they believed their role definitions were too vague. This finding led to the preparation of formal written descriptions of the co-facilitators' duties during each phase of the program. Follow-up interviews showed that making these roles explicit did much to relieve the co-facilitators' concerns and anxieties. The program developer made three presentations during the case study workshop. Word reached the program developer of comments from the women that a written summary of the presentations would solidify the content and make the presentations more valuable. In response, the program developer produced an agenda, provided written outlines of content, and listed related topics for group discussion. This example shows how formative evaluation leads to effective program changes.

2. Focus Group Meeting

The focus group is another method well suited to formative evaluation (Krueger, 1988). The focus group embodies an interactive strategy for gaining knowledge of the perceptions, experiences, and beliefs of a small group of people about a topic or experience with which they are familiar. The knowledge is generated through discussions guided by a moderator; program evaluators can make good focus group moderators. Focus groups should be flexible, relatively simple to conduct, and cost-effective. Moderators should use a group's discussions to probe clients' and/or implementers' perceptions of the strengths and weaknesses of a program. Through the focus group meeting, thorough and detailed information can be acquired that a pen-and-paper survey does not elicit. That is the advantage of focus groups, but the method has a disadvantage, too. It does not yield generalizable numbers, such as exact percentages of people holding a particular belief or encountering a particular experience. But, as has been discussed, this may not be a pertinent matter, especially during the initial implementation stage; and if necessary, focus group data collection can be augmented with a survey or other quantitative method.

An Example of a Focus Group Meeting. Quantock and Beeynon (1997) evaluated an osteoporosis awareness program using process evaluation and a focus group. The team's purpose was to see whether the program was meeting both patients' perceived needs and their medically identified needs. Sixteen female patients living within 10 miles of the hospital were asked to join the focus group. Each woman received an explanatory letter and a list of the five topics (related to the program's purpose and objectives) the focus group would discuss. Eleven patients accepted the invitation; transportation was provided as needed. The confidentiality policy for the focus group was agreed to by participants. An independent clinical psychologist facilitated the focus group; implementers of the osteoporosis program were not present. The discussion lasted about 45 minutes, after which refreshments were served.

Although data from the focus group meeting showed the program was addressing several needs the patients had, it also showed that patients desired improvement in four areas. They were dissatisfied with the staff's general level of knowledge and with the length of time required for diagnosis. They felt confident about the hormone replacement therapy they were receiving, but they were less satisfied with regard to the benefits of other employed therapies, such as biphosphonates. Above all, the focus group participants felt great fear about the future. They attributed this, in part, to what they saw as an inappropriate focus by program staff on disabilities stemming from osteoporosis rather than on the patients' remaining capacities and the potential for them to retain these capacities. Program planners implemented four changes in light of the focus group discussion: (a) The program staff's professional awareness of osteoporosis and its management was more strongly emphasized, with growth of this knowledge base becoming an ongoing requirement; (b) information provided to patients about various treatment options was made more comprehensive and equitable; and information about the benefits of treatment was given added emphasis; (c) staff members were reminded of the empowering effect of a positive attitude toward patients' future health, and realistic lifestyle advice and practical information about the risk of bone fractures were made more readily available to patients; and, finally, (d) a requirement was added to continue evaluating the program in order to address the changing needs of patients.

3. Intensive Interviews

During a formative evaluation, intensive interviews with individuals can be as helpful as focus group discussions. Face-to-face interviews of clients and/or program staff such as managers and implementers are a good means of collecting data, especially through evaluators' probing of interviewees about the more complicated issues.

An Example of Intensive Interviews. Hawe and Stickney's (1997) evaluation of one coalition provides an example of the intensive interview as a data-collection device useful in formative evaluation. This intersectorial food policy coalition had an ambitious purpose: to improve the food supply and to improve cooperation among organizations to facilitate the provision of adequate, nutritious food. After 12 months of operation, the coalition saw itself as floundering and sought an evaluation prior to trying to develop strategies to boost its productivity. It hoped for the kind of feedback that would be a catalyst for change, directing whatever restructuring of the program was needed. The coalition wanted the evaluation to be finished and the feedback information in its possession within 3 months. The evaluation literature was the basis for the researchers' decision to focus the evaluation on these areas: coalition members' perceptions about the role and responsibility of the coalition, patterns of attendance at coalition meetings, the members' degree of involvement in and satisfaction with the work, the characteristic decision-making process of the coalition, members' expectations about outcome efficacy, and members' suggestions to improve productivity. There were 21 members and former members of the coalition who agreed to be interviewed. Members were sent in advance a self-administered questionnaire regarding the coalition's effectiveness in achieving its goals.

The data that were obtained revealed a few main problems: an insufficient mechanism for attracting new members, conflicts between the perceived roles of the coalition, and a notable lack of confidence in the coalition's capacity to achieve its goals. The evaluation feedback and follow-up discussion allowed the current members to enact several changes in the coalition's operations. Summing up, the coalition's structure was recast, stronger mechanisms were created for realizing goals, and incentive management (ways to enhance benefits and lower costs to the diverse parties involved) was strengthened.

4. Comprehensive Scanning

A formative evaluation of a large program usually requires using more than one research method to acquire data in the field. This kind of formative evaluation is called *comprehensive scanning*. Comprehensive scanning rapidly identifies major implementation problems and otherwise scans for opportunities to enhance a program. Scanning differs from on-site observation and checking in two ways. The first is scale: Scanning is typically applied with large programs or programs with multiple sites operating simultaneously. The second is the method of data collection, and this grows out of the difference in scale. (Larger scale evaluations entail difficulties in using evaluators' participation or observations as a central source of data.) Scanning usually relies on simultaneous deployment of several data-collection methods, such as record reviews, telephone conferences, e-mails, site visits, interviews, and surveys. But like other formative evaluation methods, scanning is dependent on the stakeholders' program plan to guide evaluation activities. Principles discussed above (see "On-Site Observation and Checking") are thus applicable to scanning as well. The action program conceptual framework serves as a guide to the important focal areas during scanning, or as a discussion map for evaluator and stakeholders to use as they determine elements of the program plan most likely to be vulnerable and in need of close observation. Data taken from the field may raise other concerns, which evaluators should pursue if possible. Periodic forwarding of information to stakeholders allows them to take remedial action in the timeliest fashion.

Although scanning is most often needed during formative stages of program development, it is sometimes employed to ensure that a program is operating properly in its mature implementation stage. A fully mature program may be well served by a *permanent evaluation system* that monitors the primary areas of program implementation. It is important to realize that scanning is not a permanent evaluation system because the latter generally requires a great deal of time to establish and so cannot meet the time constraints of the formative stages. The permanent evaluation system is the topic of Chapter 8 of the book.

An Example of Comprehensive Scanning. The arthritis self-care project evaluated by Brunk and Goeppinger (1990) employed systematic scanning. The theory-based program plan for the project defined the intervention as a standardized curriculum teaching arthritis self-care behaviors plus problem-solving skills helpful in managing rheumatic disease. The intervention was "packaged" in two distinct modes: home study and small group. Those participants following the home-study mode completed the curriculum in their homes, overseen by trained volunteers called community coordinators. Those following the small group mode met weekly for a total of six sessions, facilitated by trained volunteers called lay leaders. Brunk and Goeppinger collected data from several sources, including audio recordings of small group sessions, weekly informal interviews with caregivers, participants' records of contact with project staff, completed worksheets, and written communication between project staff and caregivers. The information gained from the multiple methods was wide ranging and important for detecting problems for program adjustment.

Initially, project designers attempted to identify and recruit community leaders to be the lay leaders. Key informants were asked to nominate leaders, so those recruited could be trained to provide direction to caregivers and conduct the intervention. As it turned out, the evaluators learned, finding enough community leaders to serve nine scattered target areas had been staggeringly difficult. Community leaders named by key informants often were unable to join the project. As a result, volunteers had been recruited and trained to be the lay leaders.

The intervention protocol detailed in the self-care project program plan directed caregivers to present the standardized curriculum and facilitate group discussion. However, the evaluators found that actual service delivery deviated from the plan. They discovered that caregivers had skipped over topics or canceled group discussions. Remedial action taken in the face of these data included emphasizing the standardization of training, censuring caregivers' behavior, and audiotaping class sessions, all in an effort to minimize content variation.

Originally, the self-care project sought to match clients to caregivers in their home communities, maybe even to caregivers known to them. However, with the eventual pool of clients scattered across nine rural counties, that was impractical. In the end, caregivers had been assigned simply to ensure coverage rather than to create the intended pairings of caregivers with community clients.

Brunk and Goeppinger (1990) chose the term *reinvention* to indicate a change in the program plan; it is a term highlighting the positive force such a change constitutes. The researchers, who clearly encountered many changes that had taken place during development of the program, stressed the importance of systematic documentation of any changes in the intervention protocol or other area. When changes go undocumented, it is difficult to conduct a high-quality outcome evaluation and interpret its results.

Formative Evaluation Results: Use With Caution

Formative evaluation is useful when it is accepted for what it is. It is a strictly developmentally oriented approach and its results should be used only for timely fine-tuning purposes. Results of formative evaluation should never be used to describe an implementation's quality, for two reasons. First, formative evaluation has a "quick fix" nature, mandated by the needs of stakeholders when programmatic problems do surface. A problem identified by an evaluation yesterday may not be a problem today if remedial action has been taken. When conditions are so changeable, it is difficult to make meaningful value judgments. A stable pattern of implementation usually must be in evidence before quality or merit can be judged. Second, in order to provide feedback quickly, formative evaluation often must apply research methods elastically, altering certain "prefabricated" methods to suit the circumstances. The small size of survey samples discussed above is one example, and another is found in the interview with key informants. To respect stakeholders' time frame, interviews can be "tailored" to include only those key informants who are easily available. The elasticity in application of research methods creates a situation in which it would be difficult to defend the methodology if results were used to rate a program's merit. In short, by "stretching" the methodology, formative evaluation invites its designation as a less-than-rigorous approach when measured against traditional research standards. Still, the value of formative evaluation is not really diminished as long as its results are used only for program improvement. The overarching theme of this book is that the different evaluation strategies and approaches are good for different evaluation purposes: Formative evaluation may not be good for merit assessment purposes, but, by the same token, assessment-oriented evaluation may be of little use in program development.

THE PROGRAM REVIEW/DEVELOPMENT
MEETING, UNDER THE TROUBLESHOOTING STRATEGY

In an organizational setting, research is not usually regarded as the sole way to obtain information useful for making decisions about issues of program implementation. Program managers and other stakeholders often rely on organizational meetings to gather information needed to identify problems and propose solutions. Evaluators are prepared to facilitate such meetings. Adopting the troubleshooting strategy, evaluators called in during the initial implementation stage strive to foster consensus among stakeholders on implementation problems and solutions. One means of deploying this strategy is the *program review/development meeting*. Pressures of time lead to many important decisions about programs being made in ordinary meetings, unaided by the evaluator with his or her empirical field findings. Making decisions in this way is firmly discouraged by the scientific community, with its emphasis on evidence-based choices. Whatever its weaknesses, however, for the foreseeable future, the meeting convened for discussion of issues and making of decisions will continue as a modus operandi within most organizations. Interest in program development is growing among organizations, though, so the time may be ripe for evaluators to examine the meeting-based approach to decision making. Perhaps, if pursued with caution and recognizing the limitations involved, this approach could become another option available for evaluation.

The purpose of the *program review/development meeting* is to have program supervisors and implementers (or their representatives) gather, in the presence of evaluators, to talk over challenges facing their program. This approach requires the evaluator to serve as discussion facilitator *and* consultant, systematically reviewing with the stakeholders the major difficulties with the program's implementation and proposing problem-solving strategies. Discretion should certainly be exercised in the decision to use the program review/development meeting approach, which is largely good for internal utilization only. It does not have the capacity to meet accountability requirements. Stakeholders, especially funding agencies, are strongly urged not to use the program review/development meeting as a substitute for other necessary kinds of evaluations. Evaluators must communicate the limitations of the approach to stakeholders. A discussion of guidelines for using the program review/development meeting approach follows.

Program Review/Development Meeting Principles and Procedures

It is important to distinguish a program development meeting from the regular staff meetings convened by an organization. Regular staff meetings are usually conducted and controlled by a supervisor, and it is here that the limitations of such meetings start. The supervisor, having authority over the implementers, is perhaps not the first person to whom they would voice their observations about problems. It may seem too much like an acknowledgment of incompetence. Furthermore, supervisors may lack expertise in steering the discussion to systematically explore implementers' views. The risk of incomplete discussions—because of the supervisor's presence and because of the supervisor's possible deficiencies—threatens the meeting group's identification of problems and remedial actions. The program review/development meeting tries to overcome the limitations of regular staff meetings by including evaluators from outside the organization as facilitators and consultants. In a program review/development meeting, supervisors become equal partners with staff members, and the evaluator steers the discussion. Evaluators have much to contribute to such meetings.

In the role of facilitator, an evaluator helps create an open, safe atmosphere where participants freely express their opinions and recount their experiences implementing a program. For this to happen, the evaluator should be external (an evaluator who is independent from the organization and has no stake in the program) rather than internal (an evaluator regularly employed by the organization). In the role of consultant, an evaluator uses knowledge of evaluation (such as the framework of program theory) to systematically set an agenda, provide background information, and present all important issues before the meeting. The evaluator can also provide, as needed, a consultant's input concerning options to resolve problems. Prior to the program review/development meeting, each supervisor and staff member should receive from the evaluator a draft agenda intended for review and comment. Ahead of the meeting date, the evaluator should also secure from the program director and other supervisors a commitment to ensure a safe environment for discussion. The program review/development meeting should start with an announcement of the meeting's purpose and the setting of ground rules for discussion. General ground rules should be that individual opinion is honored and information from the meeting is destined for use in improving programs rather than punishing people. Again, supervisors are to be regarded as equal partners in this setting. The conceptual framework of program theory, especially the action model, is available to evaluators as a guide as they facilitate systematic discussions of the major areas of implementation and the problems therein. As remedial strategies begin to be developed in the meeting, the evaluator offers professional opinions for consideration.

Facilitating a program review/development meeting takes excellent communication and facilitation skills. Not every evaluator is constitutionally suited to the task, but disinclination or lack of skill here can be overcome by teaming the evaluator with a professional facilitator. The two work closely together to prepare the agenda and materials and serve as consultants at the meeting. The need for external evaluators has been pointed out and is strongest when stakeholders have highly divergent interests and backgrounds. In the midst of competing interests, the external evaluator tends to strike staff members as neutral and credible. (Large-scale programs, especially, benefit from using external evaluators.) Under very specific conditions, it may be possible to send internal evaluators to facilitate a development meeting:

- Good working relationships exist among supervisors and staff.
- The internal evaluators are very knowledgeable about program evaluation.
- The internal evaluators have good facilitation skills.

Internal evaluators certainly have one advantage over external evaluators: Their services are low-cost or even no-cost when supervisors agree to that arrangement.

Program Review/Development
Meeting Advantages and Disadvantages

The program review/development meeting has several advantages when it comes to providing information to enhance programs in the initial implementation stage.

- Implementers are given a sense of ownership and may enthusiastically buy into the problem identification and solution process. The meeting is an opportunity to express views and concerns about implementation and needed action. Implementers recognize that they are a real force in development, which may mean they are likelier to support proposed remedial action and other changes.

- Costs remain low. Obtaining evaluators to facilitate meetings requires some money, but much less than most other evaluation approaches requiring data collection in the field.

- Program review/development meetings produce feedback that can be turned around quickly. To write a summary report of what was learned in a development meeting takes just a few days to a few weeks, depending on what the evaluator and organization have arranged.

However, this approach also boasts a handful of significant *dis*advantages:

- Input at a development meeting can be quite impressionistic. Discussions consist largely of implementers' impressions; the accuracy or validity of impressions is not checked or verified. For example, if an implementer says that some clients are reportedly swapping program-provided food coupons for street drugs, discussion of the issue can be intensive as imaginable—but still, there is no factual information about the problem on the table. Additionally, implementation problems are usually multifaceted, whereas each individual implementer views problems from a single, personal perspective. It is not uncommon for implementers' input about problems to be partial or fragmented. Action steps coming out of the meeting then become potentially inappropriate or ineffective.

- The development meeting approach emphasizes a consensus-building process, not necessarily accuracy. To encourage participation and satisfy the meeting members, the meeting seeks parity, representativeness, and inclusiveness. Although this certainly fosters consensus, it has little to do with seeing the real problems and the optimal solutions.

- Vocal or articulate implementers may dominate a program review/ development meeting. As in most meetings, some participants in program review/development meetings will be more stirred up, more articulate, or just more comfortable speaking in public than other participants. Facilitators work to encourage those on the low end of the spectrum, but, despite their efforts, some meetings are dominated by the outspoken and the well-spoken (squeaky wheels do get the grease). Even worse, at times, the more vocal and articulate individuals bring personal agendas to a meeting and manipulate the proceedings to serve their own ends.

Example of a Program Review/Development Meeting

Gowdy and Freeman (1993) convened a development meeting as a program analysis and development tool for a program helping low-income women become and remain employed to achieve economic self-sufficiency. Services provided by this program included GED instruction; a job readiness course; employment development and placement services; and support services (transportation and child care assistance, a clothing bank, counseling, referral, and follow-up). Most clients were single mothers in their late twenties or early thirties. Gowdy and Freeman used a conceptual framework they called the *program model* (which is similar to the program theory discussed in this book) to facilitate the program development meeting. Internal evaluators were chosen because the conditions cited above for their use had been met. The evaluators began by reviewing program documents, with the program model as guide. A memorandum and copies of the program model were sent to participants in advance of the meeting, which was held at the agency and kept deliberately informal. A 3-hour meeting was planned.

Evaluators' tasks during the meeting included "translation" of the program model's meaning, purpose, and operation into the staff's operating language. Evaluators were responsible for keeping the meeting on track, as well, and creating a safe, open discussion in which all could participate. They met with success in terms of valuing each individual's opinion and avoiding domination of the meeting by any one person. As Gowdy and Freeman (1993) put it, "the receptionist's experiences of the program were listened to and considered as much as those of the director and direct service staff" (p. 69). Important findings came from this meeting pertaining to six different aspects of the program.

Implementing Organization and Implementers. Meeting participants indicated that the program staff were predominantly African-American and exclusively female, but this was consistent with the clients served. Nevertheless, hiring a Hispanic woman was suggested to increase diversity. Discussion established that staff relationships were good, but there was a need to improve attendance at and communication during weekly staff meetings. Discussion also established that meeting participants believed the training of program staff was inadequate; they felt a strong need for more new-staff orientation activities, monthly in-service training, funding to attend conferences and seminars, and tuition reimbursement. Furthermore, it was indicated that the program's resources were not enough to perform beneficial home and workplace visitation.

Target Population. Opinions expressed at the meeting were consistent about the program's clear success in targeting and providing services to vulnerable, low-income African-American women. The group believed the services should and could be expanded to benefit other vulnerable minority women, such as Hispanic and Asian women. As for potential barriers to participation, the group acknowledged that the agency was centrally located and accessible by public and private transportation. From their perspective, enrollment of clients had been no problem, thanks to active outreach and effective recruitment strategies, a flexible intake schedule, and transportation and child care assistance.

Service Delivery. In general, service delivery was adequate, according to meeting participants. If there was a weak area, they felt, it was the postgraduation period when little systematic attention was paid to clients. Despite the establishment of a goal to support women for up to 1 year while they entered the job market, no clear guide facilitated ongoing intervention once women completed classes and sought employment. Clients experienced frustration when a job did not materialize immediately, participants reported. Clients also experienced other strong emotions in this period: anxiety, excitement about learning, fear of success, and gratitude for help received. Meeting participants indicated their belief that a good deal more attention needed to be paid to clients' emotions as part of the change process.

Ecological Context. According to those present at the development meeting, the agency had a positive reputation as an advocate for minority women.

Goals and Outcomes. Participants said that goals for helping the low-income women reach economic self-sufficiency were explicit, clear, and focused on desired outcomes. They did suggest, though, that process goals, such as acquiring job-related skills, might enhance the program.

Determinants. The participants indicated that precisely how the services of the program altered the relationship of women and poverty had never been clearly understood. They identified empowerment as the determinant for achieving the goals.

Gowdy and Freeman reported that this development meeting led to significant modifications of the program. As an example, changes were made to deal with inadequacies of postgraduation follow-up. A peer support network was designed for the graduates that, in addition to providing mutual support, fostered role modeling and skill development. Additional staff were brought on board, strengthening the program's case management capacity. Staff started to touch base and maintain regular contact with clients via mailed surveys and personal outreach. Finally, a committee was named to review and assess progress in light of the changes.

This example demonstrates that information from the program review/ development meeting is useful to internal audiences *if the meeting has been approached with caution.* The information, however, cannot be used as evidence of a program's quality or effectiveness for an external audience because the approach cannot ensure the credibility of the information.

BILATERAL EMPOWERMENT EVALUATION, UNDER THE DEVELOPMENT PARTNERSHIP STRATEGY

Evaluation activities are ever more diverse, and program stakeholders have started inviting program evaluators to collaborate as their partners in the development (designing and implementing) of programs. When such a development partnership is formed, evaluators become members of the program development team and part of the decision-making process. The taxonomy refers to this as the development partnership strategy. Evaluators using this strategy will negotiate points with key stakeholders, but, more importantly, it will empower them to put evaluation results to work to devise or strengthen programs. The latter is not by any means an easy job, requiring sustained effort from evaluators; but the development partnership strategy is one available tool.

An evaluation approach associated with the development partnership strategy is the *bilateral empowerment evaluation.* Bilateral empowerment results when evaluators welcome stakeholders to join the evaluation process, and stakeholders welcome evaluators to join the program development process. Bilateral empowerment evaluation has gained strong momentum in the evaluation of community coalition programs. Whether an "alliance," "consortium," "partnership," or "network," the general aim of this type of organization is to empower a community by building its capacity to solve community problems. Funds for community coalition programs frequently lack well-specified requirements, allowing the community freedom to develop strategies and take actions as it sees fit. Furthermore, the goals of programs like these tend to be broad and ambitious; they might call for reducing drug abuse, or ending discrimination, throughout an entire community. Evaluating community coalition programs is necessarily a complex and dynamic process because the work proceeds on two fronts—the evaluation process and the program development process.

Evaluation Process

The evaluative procedures of the bilateral empowerment evaluation encourage sharing. The evaluation process is a mutual learning experience for stakeholders and evaluators, and each has empowered the other to improve the program. Evaluators also are obligated to give technical assistance to stakeholders and to build their capacity to weigh important issues wisely. In the bilateral empowerment relationship, joint decision making is the norm, extending to tasks such as conceptualizing community development and devising instruments to measure it, or gathering pertinent data.

Example of Bilateral Empowerment Evaluation

A good illustration of bilateral empowerment evaluation is found in the evaluation of a community prevention alliance conducted by Goodman et al. (1996). The alliance proposed to work for change in a community's norms regarding the use of tobacco products, alcohol, and street drugs. It was hoped that their efforts would (a) alleviate problems in the community's workplaces concerning use of these items, (b) reduce addiction and violence among 12- to 17-year-olds in the community, and (c) limit the annual incidence of HIV/ AIDS and other sexually transmitted diseases in the community. The stakeholders had worked closely with evaluators to plan, develop, and implement the intervention program. Goodman et al. (1996) broke down the process into "phases" of work, as follows.

Phase I: Program Formation. The very first need was to create the evaluation coalition by hiring staff and recruiting members from across the community. Then, the coalition was involved in carrying out a needs assessment, data from which would inform the determination of its intervention strategies. In this phase, evaluators serving on the coalition assisted stakeholders in completing a logic model of the intervention and developing indicators to measure the progress of program development. The evaluators attended coalition meetings to debrief the membership and discuss the latest evaluation results. They continually monitored the work on the program plan and encouraged stakeholders to fine-tune both systematically and as needed to ensure that the implementation would ultimately remain consistent with the plan. For instance, they monitored the effectiveness of coalition meetings because the organization of the new coalition depended on that effectiveness. When attendance and participation at meetings were both found to be lacking, evaluators brought the relevant data to the coalition members for discussion, leading to a strategy to improve rates of attendance and participation.

Phase II: Plan Implementation. Eventually, the time came to implement the strategies approved by the coalition: awareness campaigns, service programs, and policy initiatives. Evaluators began to monitor the level of effort put forth by the coalition in pursuing each of these. They found evidence of a strong emphasis on awareness campaigns, one that detracted from the implementation of other activities. In addition, evaluators felt that awareness campaigns were not likely to create lasting change. Evaluators turned to the evaluative data to interest coalition members in refining the intervention to better stimulate the community to change. The evaluators also worked to build the coalition's capacity to implement the program plan effectively and to increase its ability to use evaluation results well.

Phase III: Impact. The final phase featured efforts to institutionalize coalition strategies, determine what the communitywide results of the intervention were, and preserve the coalition following termination of the grant. A successfully implemented coalition could, it was thought, boost community awareness of, concern for, and action on substance abuse, violence, HIV/AIDS and other STDs, and teen pregnancy. Goodman et al. (1996) suggested such research methods as a survey of key community leaders, a broader community survey, and trend analysis for use in the assessment; however, the actual performance of the impact assessment was beyond the scope of their study.

About the Evaluators' Role. The essence of the evaluator's role in program planning and development under the bilateral empowerment evaluation approach happens is expressed well in Goodman et al. (1996):

> In general, we stress to Alliance members that the best ways in which we [evaluators] can be helpful are by being dedicated to an ongoing relationship with the coalition and by providing honest feedback that is based on data, open sharing of information, problem solving, negotiation, good will, and support of the coalition's effort. While the coalitions that we evaluate do not always agree with our approaches, conclusions, or recommendations, they view us as valued members who provide important feedback. Without earning the trust of our community coalition through open communication, negotiation, and compromise, we do not believe that our assessment approach is feasible. (p. 58)

Pros and Cons of Bilateral Empowerment Evaluation

In a fluid, complex program such as a community coalition or consortium, there is often recognition that evaluation information will be crucial for directing the program's development. Stakeholders having this mind-set tend to invite evaluators to partner with them in the planning and program development process. This participatory quality is seen again in bilateral empowerment evaluation, in which evaluators move beyond formative evaluation to serve as planning and development team members. They work closely with stakeholders to select an intervention, structure the program, and solve implementation problems.

Bilateral empowerment evaluation, like any other approach or strategy, has its pros and cons. Its chief advantage is that it maximizes the impact of evaluation data on the decision-making process. Bilateral empowerment allows evaluators to apply the development-oriented process evaluation discussed in this chapter. Later on, however, these evaluators are likely to need to bow out of assessment-oriented evaluation concerning the merit of the program they helped launch. Because of the risk of conflict of interest, the credibility of a merit assessment carried out by evaluators using the bilateral empowerment strategy would, in a broad context, be compromised.

SEVEN

ASSESSING IMPLEMENTATION IN THE MATURE IMPLEMENTATION STAGE

Programs in the mature implementation stage are those in which the procedures and rules of implementation have become routine. An assessment of how a program is implemented in the mature stage is called *process evaluation*. There are two kind of process evaluation. At the mature stage, stakeholders could ask evaluators to conduct an evaluation to identify implementation problems in a timely manner, as discussed in the previous chapter. This kind of evaluation is called a *development-oriented process evaluation*. However, more often stakeholders ask evaluators to assess how well the program was implemented, which is called an *assessment-oriented process evaluation*. This chapter starts with a brief review of these two types of evaluation and then moves to a discussion of assessment-oriented process evaluation.

DEVELOPMENT-ORIENTED AND ASSESSMENT-ORIENTED APPROACHES

Even a matured program may be subject to things of a programmatic nature going wrong. When this happens, stakeholders may want evaluators to assess the problem quickly and provide information to help them create a remedial strategy. Thus, those previously introduced (see Chapters 4-6) development strategies and approaches that stress timeliness can also be used in evaluation at the mature implementation stage. The various strategies can be just the tools needed to identify stakeholders' development needs and pinpoint the corresponding evaluation techniques. Despite its potential usefulness, development-oriented evaluation is used less frequently at the mature implementation stage than is assessment-oriented evaluation because the program usually needs to be evaluated to meet accountability needs. At the mature implementation stage, concern for accountability begins to grow among external stakeholders, such as funding agencies and decision makers. Internal stakeholders, too, become inquisitive about accountability. Generally speaking, development-oriented evaluation can be carried out by either internal or external evaluators. However, assessment-oriented evaluation is preferably done by external evaluators. This chapter's primary topic is assessment-oriented process evaluation. First, however, is a brief discussion of development-oriented process evaluation used in the mature implementation stage.

APPLICATION OF DEVELOPMENT-ORIENTED PROCESS EVALUATION IN THE MATURE IMPLEMENTATION STAGE

Development-oriented process evaluation can be very productive in the mature implementation stage when a program rationale or program plan needs to be revisited for clarification or modification. Development-oriented approaches are also useful when implementation problems call for some troubleshooting.

When Stakeholders Need to Modify or Clarify a Program Rationale and Plan

At times in a long-running program, stakeholders such as the program director, implementers, and decision makers can come to feel that they have lost sight of the goals being pursued or the direction in which the program is moving. They want to revisit their program rationale and program plan to clarify or even reshape the program. Development-oriented strategies are useful here, and an especially popular one is *development facilitation*. Many consensus-building tools are available within program evaluation that can facilitate the work of stakeholders needing to redefine their goals or reprioritize a set of existing goals. When prioritizing goals is program stakeholders' and evaluators' main interest, traditional techniques—for example, the Delphi method, Nominal Group, and Multiattribute Utilities Method—can serve the purpose (see discussions in Chen, 1990). However, when evaluators and stakeholders are pursuing issues that affect relationships among program objectives and goals, they will find more help with the *conceptualization facilitation* and *concept mapping* approaches (see the taxonomy, Table 3.1). Both of these techniques are related to program theory, but each offers a unique advantage. Concept mapping, a quantitative tool, is especially helpful when a program has numerous goals and activities; conceptualization facilitation is appropriate when relationships among components of the rationale and plan need to be identified or made clearer (as elaborated in detail in Chapters 4 and 5).

Conceptualization Facilitation Approach

Facilitating stakeholders' efforts to elucidate their program rationale and program plan should involve consensus building, so the working group format of the conceptualization facilitation approach is preferred to intensive interviewing. The conceptualization facilitation approach works systematically, so it can be employed to develop rationales and plans as well as improve them. The evaluator engaged to help with either task will find in Chapters 4 and 5 much guidance for formulating evaluation designs and procedures of a conceptualization facilitation type.

Concept Mapping Approach

Concept mapping is a quantitative tool with which stakeholders can clarify program objectives and identify relationships among them (Trochim & Cook, 1992). The approach is a structured process soliciting the opinions of various representatives of stakeholder groups. Ideas for the program contributed by these representatives are organized in a graphic representation called a *concept map.* The concept map, then, illustrates the stakeholders' theory of the program's basic components and their interrelationships. Procedures for applying concept mapping are illustrated in the following example.

An Example of Concept Mapping. A study of a youth drop-in center by Mercier, Piat, Peladeau, and Dagenais (2000) provides an example of the application of concept mapping. The YMCA drop-in center stated its purpose as offering 10- to 17-year-olds an informal setting for unstructured and structured activities after school hours and on Saturdays. Structured activities included sports and recreation programs, educational and sensitization programs, and informal counseling and referral services. This YMCA center saw itself as an alternative to unsupervised settings and a deterrent to undesirable behavior such as substance use, intergenerational and ethnic conflict, sexual activity and resulting pregnancy, and dropping out of school. The board members and staff of the drop-in center had expressed a need to better understand the preventive nature of the center to make their strategic planning more effective.

The center's director and three staff members participated actively in devising a concept map. At their initial meeting with evaluators, they were asked the general question, "In what way can/does the youth center contribute to prevention?" They brainstormed answers in the form of short statements, and a total of 98 statements were generated. In a subsequent meeting, each participant was asked to rank the importance of each statement relative to the others and to sort the statements into thematic groups. This yielded, for the concept map, the basic groupings or components of the program. At a follow-up meeting, participants were asked to identify causal relationships among these components. Data analysis was then conducted and three desired outcomes for the program identified by staff: (a) to offer youth an alternative to the street, school, or family; (b) to promote personal and social development; and (c) to sustain leadership development. Staff members believed that these outcomes were pursued through such means as the flexibility of center program and activities, special events, the freedom to experiment while supported by supervision, recognition of achievement leading to a sense of self-worth, and so on.

When Stakeholders Need to
Troubleshoot Implementation Problems

Stakeholders may not always be satisfied with an implementation process, even when their program has reached the mature implementation stage. They may invite evaluators to look into perceived problem areas and generate relevant information helpful to them as they consider their next move. In such cases, troubleshooting strategies and approaches should meet stakeholders' needs. (These approaches are discussed thoroughly in Chapter 6.) A study of a health education program for pregnant women in Chile (Foster, 1973) provides an interesting example of development-oriented evaluation at the mature implementation stage. Public health centers there arranged for women who had just been informed that they were pregnant to attend prenatal classes. The program was modeled on an American program quite popular and successful in the United States. In Chile, however, its success had been limited, chiefly because expectant mothers failed to attend classes. Research showed a problem with an aspect of the program's service delivery mode—its setting. The targeted women objected to being taught in classrooms like children. Prompted by this finding, public health decision makers restructured the classes as "clubs," which typically met in the women's various homes. In Chile, club membership connotes middle- to upper-class status, so the women then enjoyed participating. The program contributed refreshments for each meeting, and the intervention became a social affair where conversation was mixed with exchanges about prenatal care. Changing the service delivery mode led to a quite successful program.

ASSESSMENT-ORIENTED PROCESS EVALUATION

Development-oriented evaluation strategies and approaches are primarily for internal use to address problems in a program immediately (see Chapter 5). Their results do not provide sufficient information about how well a program is implemented. When external stakeholders (funding agencies, decision makers) and/or internal stakeholders want to know how well a program is being implemented, an assessment-oriented process evaluation is called for. The conceptual framework of an assessment-oriented process evaluation is illustrated in Figure 7.1.

Figure 7.1 indicates that this kind of evaluation is an assessment of the actual implementation process compared to its original program plan and rationale.

Assessment-oriented process evaluation serves one or more of the following three purposes.

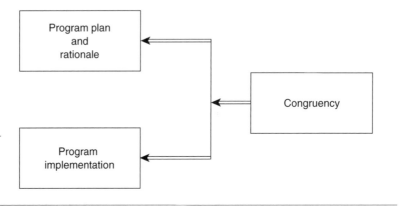

Figure 7.1   Conceptual Framework of an Assessment-Oriented Process
                    Evaluation

Meeting Accountability Needs

There is a tendency to think of accountability as meeting (or falling
short of) program goals. In actuality, program outcomes comprise just one
area of accountability. Another area of accountability that is of considerable
importance is the program implementation. Funding agencies, decision makers,
and other external stakeholders are very interested in process questions, such
as, Who is being served by the program as it is implemented? and, What is the
quality of the services being provided? These and similar questions can be
answered through process evaluation. The results of process evaluation will
also, of course, be of interest to internal stakeholders (program directors,
implementers) who have much invested in the quality of implementation.

Systematically Meeting Program Improvement Needs

A basic process evaluation can provide information on whether an
intervention is implemented as intended. However, an advanced process such
as theory-driven process evaluation can also answer how the implementation
of each element has contributed to the overall quality of the program. This
kind of information is very useful for understanding the general direction
in which a program is moving and whether there is a need to modify the
program plan.

Providing a Context for Interpreting Outcome Evaluation

An *outcome evaluation* may tell whether or not an intervention indeed affects an outcome. However, evaluation findings concerning the relationship of intervention to outcome per se are difficult to interpret or use in the absence of contextual information about the implementation. For example, what is signified when outcome evaluation shows no relationship between intervention and outcome? Is it that the intervention is inappropriate? Or that there are serious flaws in the implementation process? The data from process evaluation generate the needed contextual information and thus can foster useful interpretation of outcome evaluation results. For example, if a program produces none of its expected outcomes, process evaluation will be able to provide information about the source of the failure: Is it due to erroneous conceptualization of the program, or to some deficiency in the implementation?

ISSUES IN APPLYING
ASSESSMENT-ORIENTED
PROCESS EVALUATION IN THE
MATURE IMPLEMENTATION STAGE

Conducting a Preprocess Evaluation

Unlike development-oriented evaluation, which can be applied to literally any program, the meaningful application of assessment-oriented process evaluation is more restricted. Simply being implemented over some period of time will not necessarily make a program a candidate for this kind of process evaluation. For example, the money and resources designated for a program might have been usurped for activities irrelevant to that program (Rossi et al., 2004), leaving something of a vacuum for the evaluator. Process evaluation is more expensive than development-oriented evaluation, and attempting it when a program appears unprepared for it can be a disservice to stakeholders. Of course, when stakeholders insist, process evaluation can be performed in less-than-desirable evaluation circumstances. In general, though, before launching a process evaluation, it is a good idea for the evaluator to check whether a program meets certain criteria associated with meaningful assessment-oriented process evaluation. These criteria—which speak to the feasibility of providing a program with process evaluation—are the following:

1. *Resources with which to implement the program have been committed by an organization.* To conduct a meaningful process evaluation, evidence should exist that the necessary resources and staff have been assigned to implement the program. The presence of a program plan does not always mean that the program plan has been implemented. However, when stakeholders are concerned that a program's resources have been misdirected, or when they ask directly *why* a program is not effective, then process evaluation may be appropriate. Otherwise, unless an implementing organization has formally committed resources with which to implement a program, money is sure to be wasted conducting process evaluation.

2. *Stakeholders have a clear idea of what they intend to implement.* As we have seen, a general purpose of process evaluation is to check for congruency between a program as implemented and that program as intended. Stakeholders' intentions for their program must be known before the evaluator can begin such an evaluation. If stakeholders cannot communicate to the evaluator the intentions of their implementation, there is simply no way to do process evaluation. The formal program plan is the best choice for a document establishing a program's intentions. When no program plan is available, the evaluator should clarify the stakeholders' intent by reviewing documents (e.g., grant proposals) and interviewing the stakeholders.

3. *Sufficient time has elapsed for the implementation to mature.* A certain period of time is needed for a program to become mature, and only a mature program is ready for process evaluation. Conducting this kind of evaluation with a program in the initial implementation stage, or one that is still fluid, may result in the unfair conclusion that implementation of the program is poor. A rule of thumb for gauging maturity is that the larger a program is, the more time it needs to mature.

It should not be assumed that these prerequisites are easy to meet. In a worst-case scenario, evaluators might learn that the organization has difficulty articulating the exact nature of its intervention—that is, what it will do upon being funded—even though the organization has received funding for some years! In such a case, it is not possible to meaningfully assess the quality of implementation. Such a shortcoming in program design and implementation is sometimes the result of using an outside professional grant writer whose proposal looks good on paper but does not speak to the people on the inside.

Sometimes, implementation falls short because the person who developed the proposal has left the organization and others did not have the skills or information to continue the project. (High rates of staff turnover are one major weakness of service delivery across private and public sectors.)

The evaluator invited to carry out an assessment-oriented process evaluation needs to conduct a preprocess evaluation to determine whether the program meets all three prerequisites above. If it does meet them, if process evaluation is feasible, then the evaluation can proceed. If the criteria are unmet, however, the evaluator should discuss with stakeholders two evaluation alternatives. The first is to press on, in order to assess in detail the mismanagement of the program. The second alternative, perhaps more meaningful and constructive, is to initiate a development-oriented implementation evaluation (discussed in Chapters 4-6) to help the program regain the right track. At the least, stakeholders need to be made aware that there are strategies and approaches better suited than process evaluations when program rationales and plans need attention (see Chapters 4 and 5). The stakeholders should understand that the alternative strategies and approaches can actually establish feasibility for an evaluation of program implementation.

Know Which Kind of Assessment-Oriented
Process Evaluation Fits Stakeholders' Needs

As indicated in Figure 7.1, a common tactic of an assessment-oriented process evaluation is to conceive of the quality of a program's components as a congruency between stakeholders' intentions for the program (their program plan) and the facts of the program's actual implementation. Congruency between program plan and program implementation is widely understood to signify a good-quality implementation.

Two kinds of assessment-oriented process evaluation will be discussed in the rest of this chapter: fidelity evaluation and theory-driven process evaluation. Fidelity evaluation questions whether the intervention has been implemented as intended and/or whether it is serving the intended target groups. Basic accountability needs can be met through application of this evaluation. Theory-driven process evaluation is a comprehensive, systematic assessment of the implementation of all major components of a program plan.

In conducting assessment-oriented process evaluation, evaluators always need, first and foremost, to understand clearly stakeholders' needs and the evaluation strategy approaches most adequate to those needs. The remainder of this chapter tries to contribute to such understanding by detailing both fidelity evaluation and theory-driven process evaluation.

Clarify Program Intentions

The evaluator should have a clear idea of the program's intentions even before choosing his or her assessment-oriented evaluation strategy. When the intentions are plainly set forth in a well-developed program plan, evaluation of this kind is straightforward. Clarification of intent, in such a case, is accomplished mainly by discussing the program plan with stakeholders, ensuring it is up-to-date and reflects their view. But if there is no sound, formal program plan available, the evaluator needs to compile and consult what documents do exist and meet at greater length with key stakeholders. Performance assessment strategy emphasizes target groups and intervention components. Usually, agreement concerning clarification of intent for the proposed intervention and target groups can be reached following one or two meetings with key stakeholders. If the stakeholders are interested in enlightenment strategy, however, a few meetings are likely to be necessary to identify the major components of the program. Chapter 5 presents principles and procedures for creating a program plan, and these make a very useful guide to identifying these major components. (Involving the evaluator as a facilitator during the development of the program plan cuts the amount of time spent clarifying intentions.)

Occasionally during work to clarify stakeholders' intentions for their program, the evaluator runs into factions at odds over certain program components. Imagine, for example, an HIV counseling and testing program in which program managers and officers at the federal level intend to restrict services to high-risk people; but implementers at the state level intend to offer some response to anybody who walks into their clinics. Ideally, consensus on a program's intentions precedes evaluation, but if only a few points are disputed, the evaluation can proceed. The evaluator can actually use the evaluation as an opportunity to collect empirical data—findings from the field—that may give the opposing stakeholders what they need to resolve their differences. Both federal HIV program officials and state HIV implementers would, perhaps, appreciate knowing that data show overly restrictive clinics to discourage both high- and low-risk people from visiting.

Select Research Methods

Assessment-oriented process evaluation attempts to provide credible information about the quality of implementation. Naturally, such an exacting review requires time—a good deal more than development-oriented evaluation requires. The methodology of assessment-oriented process evaluation has to be rigorous, and the evidence has to be convincing. Flexibility of research methods is not desirable here, as it is for development-oriented process evaluation. Whether the research method chosen for the particular evaluation is a qualitative or quantitative method, the evaluation needs to follow closely the prescribed research procedures and requirements. Because emphasis is *not* placed on speedy return of feedback, any proposed change to a normative research method requires good justification. For example, assessment-oriented process evaluation conducted with quantitative research methods demands that issues such as sampling, sample size, measurements, significance tests, and so on be addressed. A further departure from standard development-oriented evaluation practices is that the written summary of a process evaluation may be submitted to stakeholders as a complete package following the conclusion of research. (Remember that development-oriented evaluations tend to provide feedback quickly regarding any indication of a major problem. They often do not require a formal and complete report, as do those in assessment-oriented process evaluation.)

THE FIDELITY EVALUATION APPROACH, UNDER THE PERFORMANCE ASSESSMENT STRATEGY

The taxonomy (Table 3.1) shows that the *fidelity evaluation approach* is the major approach associated with performance assessment in the mature implementation stage. Fidelity evaluation is process evaluation that gauges the degree of congruency between intervention and target groups as planned and intervention and target groups as implemented. Many fidelity evaluations (as shown below) pursue one issue: the assessment of whatever element of the action model conceptual framework is of special interest to stakeholders. Any of the four popular types of fidelity evaluation about to be introduced here can serve that need, depending on which element of the model will be scrutinized. The four types are *intervention fidelity evaluation, referral fidelity evaluation, service delivery fidelity evaluation,* and *target population fidelity evaluation.* Chen (1990) covers several other types of fidelity evaluation associated with the remaining program components, including contextual support and organizational collaboration, and provides relevant evaluation strategies and examples.

Intervention Fidelity Evaluation

Evaluators have used the terms *intervention fidelity* and *treatment integrity* to refer to what in this book is called *intervention fidelity evaluation.* The point of intervention fidelity evaluation is to see if an intervention implemented in the field is turning out as patterned by the original program plan. Generally speaking, implementations that reflect most clearly the intent expressed in the program plan are those of higher quality, and thus those most likely to work. To conduct intervention fidelity evaluation, stakeholders and evaluators must first identify the crucial elements and strengths of the intervention as it was *intended* to function. An intervention often consists of a number of elements. An antibullying intervention in a school system, for example, might include the following elements enumerated in the associated program plan: adopt school policies on bullying; hold training and discussion sessions for students, staff, and parents; obtain additional play equipment; and utilize adult supervision. Evaluators need to work with stakeholders to identify the *crucial* elements of the intervention so that they can be included in the assessment.

An intervention's fidelity to a program plan can be assessed using one or more of the following measures: coverage, strength, and intensity. Measuring coverage means asking whether the real-life implementation covers all crucial activities of the intervention as prescribed during planning. Measuring strength means determining if the implementation includes as much of the intervention, per session, as planners intended; that is, is the prescribed "dose" of the intervention being administered at one time? Measuring intensity means counting the number of sessions, or times the intervention was carried out, once again to see if the prescribed number or dose was provided. For instance, training sessions within the school antibullying program could be assessed in terms of whether they covered adequately the prescribed antibullying topics, the number of minutes they lasted, or the total number of sessions offered to students.

Example of an Intervention Fidelity Evaluation. An intervention fidelity evaluation was conducted on a school-based nutrition education program in

Georgia (Davis et al., 2000). The program was established to help fourth- and fifth-grade students consume more fruit, fruit juice, and vegetables (nicknamed "FJV") each day. The intervention in this program had three major elements: its curriculum, family activities, and point-of-purchase activities. Twelve sessions were worked into the curriculum, each lasting 40 to 50 minutes. Teachers presented the curriculum, which comprised various tasks classed in categories such as affect (increasing students' enjoyment of eating FJV) and asking skills (enhancing students' ability to ask for FJV at home). Researchers observed classrooms in order to gather data about the fidelity of the curriculum's implementation. Parents and families of the students were also drawn into the program via homework assignments and videotapes about FJV. To gather data about the fidelity of implementation of FJV family activities, evaluators used telephone interviews with parents to ascertain whether and how many homework assignments and videotapes had come home, and to ask if the family had joined in homework activities, viewed videotapes, and/or set FJV goals. During these telephone interviews, parents were also asked about participation in point-of-purchase activities held in the evenings at local grocery stores. The additional questions, plus evaluators' observation of the in-store sessions, generated the fidelity data concerning this third element of the implementation. Data analysis indicated that, overall, elements of the program had not been implemented as designed. Teachers failed to deliver the entire curriculum, selectively underimplementing messages that were crucial to promoting behavioral change. Few families attended evening point-of-purchase sessions. At-home family participation was modest, declining substantially between the fourth grade and the fifth. More than one-third of parents of the fifth graders said they did not participate in any of the child's FJV homework activities; about the same number said they had not received a videotape.

Referral Fidelity Evaluation

The fidelity evaluation approach is also used to assess the adequacy of the referral process of a program. This kind of evaluation is particularly beneficial to programs serving populations whose behavior is high risk, such as drug abusers. Clients enrolling in an intervention for high-risk behavior usually have multiple problems. For example, a person admitted to a mental health treatment program may be dealing with alcoholism, homelessness, hunger, and more. For this person, the success of the mental health treatment depends not just on the intervention but also on the alleviation of the other problems. Thus, the referral network with other programs that serve the individual's additional needs is vital to the success of the mental health intervention. The mental health program could easily fail unless it recognizes related barriers to the client's well-being and knows who can help bring these barriers down. The chief question asked by referral fidelity evaluation is, "Is there a referral process in this program that is actually functional?"

Example of a Referral Fidelity Evaluation. An example of evaluating referral services for program clients is found in Marx, Hirozawa, Chu, Bolan, and Katz (1999), an article about clients needing referrals from an HIV testing/ counseling program to prevention services. The program, also known as Counseling, Testing, Referral, and Partner Notification (CTRPN), is the largest standardized HIV prevention effort in California, but it is a brief one, generally encompassing only two sessions. To serve clients effectively, CTRPN needed to provide further attention to high-risk individuals by referring them for additional prevention services. Its guidelines clearly indicated that one of CTRPN's major objectives was to "provide referrals to HIV positive and high risk HIV negative persons for necessary medical, preventive and psychosocial services."

Data for the study of CTRPN's referral fidelity included information from the San Francisco Department of Public Health and Municipal STD Clinic. The research showed that the overall freelance rate of referral by CTRPN was low: A referral had been received by 19.1% of the health department sample and by 10.6% of the clinic sample. The authors concluded that opportunities were being missed to link high-risk clients who had been tested or counseled to additional HIV prevention services outside CTRPN. They urged that the referral component of the HIV counseling/testing program be improved.

Service Delivery Fidelity Evaluation

Service delivery fidelity evaluation ascertains congruency between the setting, mode, and procedure of service delivery as planned and as actually manifested. Service delivery is one part of an intervention that inarguably should be assessed. Service delivery fidelity is especially crucial for intensive programs like those often found in the fields of mental health, alcoholism, and substance abuse. Intensive programs often use sequential steps or sessions to complete service delivery. Serving clients properly involves steps such as intake, screening, risk assessment, case management, client readiness, treatment sessions, and so on. The success of a later step depends on how appropriately the earlier step or steps are being implemented, as seen in this example from an alcoholism treatment center. When admitted to a treatment program, the alcohol-addicted individual must be immediately assessed as to whether detoxification will be needed before treatment begins. If detoxification is ruled out, the next step is the determination by staff of the treatment that is most appropriate. Appropriate screening and diagnosis early on helps ensure that clients receive an effective treatment in a timely manner. Appropriate procedures for screening and diagnosing newly admitted clients are usually well documented in a program's operating manuals and other guidelines. Service delivery fidelity evaluation can be used to explore whether an intervention is being implemented in keeping with all prescribed procedures and rules.

Example of Service Delivery Fidelity Evaluation. Further illustrating service delivery fidelity evaluation is a Taiwanese project bringing middle-school dropouts back to the classroom. One intervention used by the project was teacher counseling of dropouts and their parents together, encouraging return to school. The counseling was delivered via regularly scheduled home visits. In 1993, I participated in a process evaluation of the project, through which we eventually learned that teachers were making very few of these home visits. The collected data showed that meeting with families was a much harder task than program designers ever imagined. Not only did busy working parents have little time for meeting teachers during the day, teachers (most female) did not feel safe visiting in distant—and, often, comparatively crime-ridden—neighborhoods at night. Beyond such logistical stumbling blocks, many parents also felt that they lacked much control over their children's behavior; they felt they could not, for instance, guarantee their children's presence during home visits. The completed service delivery fidelity evaluation indicated that the home visits were failing as a service delivery mode for this particular project.

Target Population Fidelity Evaluation

Target population fidelity evaluation looks at programs' contact with their target populations. Programs must reach sufficient numbers of clients from the specified target population in order to be effective. It might seem safe to assume that any implementing organization would maintain records securing this important information, but many, unfortunately, do lack good systems for client data. This is especially true of programs working with street outreach, capacity building, and similar techniques. Street outreach service programs almost always have information about how many hours outreach workers spend on the streets in a day, and where. However, they seldom gather information on how many contacts were made or who signed on as a client. Similarly, a program providing capacity-building services to community organizations may maintain information on how many calls are fielded per day, yet have no data on which organizations called, the nature of the calls, or the services ultimately provided. An evaluator having the opportunity to become involved in early stages of program development may be able to facilitate the establishment of a data-collection system for recording client information. If this involvement has not occurred, however, and if implementation began long ago without a data-collection system, other means for estimating how many target population members are served are available. In the case of the street outreach program, for instance, a survey of a representative sample of residents in the targeted areas could be used to estimate how many people are being contacted by the outreach workers. In the capacity-building case, evaluators could contact those community organizations that have telephoned the program over a specific period and talk with them to reconstruct the purposes of the calls and the services provided. To conduct a target population fidelity evaluation, evaluators need to ask three main questions:

1. *How many clients were served by the program during a specific period?* Counting the clients served may or may not be a straightforward prospect. The evaluator must remain aware of the distinction between clients recruited and clients served, although this distinction is not much of an issue for programs delivering their services without delay. Through street outreach, services can be provided on the spot and clients counted straightforwardly. The same is true of a hotline program: Count the number of people who make calls that get answered. In such cases, the intervention follows on the heels of recruitment and is soon completed. With intensive programs, however, and as discussed above, weeks or even months may be needed to complete services, and clients must be tracked over that period in order to be counted accurately.

2. *How many of the clients served come from the target population?* It is not necessarily true for all programs that all those served come from the intended target population. Implementers tend to want to serve clients who are easily accessible. As an example, consider a program intended to entice school dropouts back to the campus. Program implementers, recognizing that people who have left school prematurely are often difficult to reach, gradually refocus on serving current students with a high risk of dropping out. Any inability to focus earmarked resources on the target population is usually of great concern to external stakeholders.

3. *Does the number of clients served justify the program's existence?* Upon determining how many clients served come from the target population, the evaluator's next question elicits a judgment call about a program's performance, based on that number. Has the program served enough clients? This question is easy to answer *if* projected numbers were included in the program plan or other document. Some programs have a clear policy for client numbers during a period of time, such as a child care program requiring the enrollment of at least 20 children of welfare mothers for daily care. In the absence of a clear criterion, the evaluator will need to consult with stakeholders about the issue before beginning the evaluation. Stakeholders usually can cite at least a general count considered an acceptable client base. Alternatively, the evaluator can use average numbers from comparable programs as a basis for determining if the program is used enough. Insufficient numbers can lead to a declaration of the program's failure. For example, according to its program plan, a drug abuse prevention program expected to pull in a few hundred youngsters with after-school, neighborhood-based activities. Just a few children turned out. It is not hard to view these numbers as a sign that the program failed. The failure of a program is surmised if the program cannot reach a sufficient number of clients within the target population, cannot adequately screen them for appropriate intervention, or cannot retain most clients throughout the process.

An Example of Target Population Fidelity Evaluation. Among evaluators, a popular source for data about clients served is the records that implementing organizations keep. Glasgow, Lando, Hollis, McRae, and La Chance (1993) used client contact records to evaluate the reach of a smoker's hotline that provided a variety of smoking cessation services and was free to more than 2,100 HMO members. The program was well promoted via a variety of channels, including newsletters and other mailings. Hotline scripts were pilot tested, and experienced telephone staff and smoking cessation counselors were trained. During 33 months of operation, however, only 305 calls came in to the hotline, according to program records. That was an average of 2.3 calls per week (with 71% of all calls coming in just after the program started), meaning the cost of each call to the hotline was an estimated $81. The program was regarded as a failure because it was not reaching clients.

Issues on Fidelity Versus "Reinvention" in Process Evaluation

The fidelity evaluation approach is predicated on the idea that fidelity of implemented program components to intended components is highly desirable. It also assumes that less-than-complete fidelity reduces the effectiveness of programs in the field. These beliefs are not without controversy. There is within program evaluation a school of thought called the *diffusion tradition,* which argues that change is a necessary part of the adoption of any program (see Blakely et al., 1987). Such change is termed *reinvention;* its occurrence is, according to its proponents, absolutely necessary to preserve program effectiveness. Thus, discrepancy between a program plan and the observable implementation of the plan is *desirable* and should be encouraged.

Fidelity evaluation and the diffusion tradition are opposing viewpoints, but they can be somewhat reconciled by the taxonomy and the contingency perspective as presented in this book. Reconciliation is accomplished in this way: In the initial implementation stage, a program needs frequent adjustment as it fits itself to local circumstances; so reinvention is a *good,* one which can be facilitated by development-oriented process evaluation. Later, when a program is in the mature implementation stage or has operated in the field for a long period, making modifications can inhibit the smooth, efficient operation of that program. Unless an undeniable reason for changing the program crops up, during the mature implementation stage, the fidelity of the implementation to the intention should be protected. Implementation deviating too much from the program plan can dilute the program's integrity. Thus, in the mature implementation stage, assessment-oriented process evaluation is appropriate, just as development-oriented process evaluation was appropriate in the earlier stage. Given the reconciliation that the taxonomy makes possible, it is not required that no change ever occur in the mature implementation stage. It is, however, assumed that stakeholders can justify a change they make. If an HIV prevention program decided its target population would no longer be gay white men but instead would be gay African-Americans or Hispanics, the change would be justified by the shifting demographics of the infection. Funding agencies would not look upon such a change as being arbitrary, but it would remain important to document the change, and any others in the implementation, for accountability purposes.

ENLIGHTENMENT ASSESSMENT STRATEGY AND THEORY-DRIVEN PROCESS EVALUATION

The fidelity evaluation approach found under the performance assessment strategy typically takes on the assessment of just one or two action model components. However, in some situations, stakeholders may seek a complete assessment of the whole implementation of a program. *Enlightenment assessment strategy* fits this bill. Enlightenment assessment is an undertaking that can help determine the overall quality of implementation and fosters a systematic formulating of strategies—as opposed to piecemeal reacting—meant to improve the implementation. The evaluation approach associated with enlightenment assessment is called *theory-driven process evaluation* (see Table 3.1 for its place in the taxonomy). Theory-driven process evaluation systematically assesses how the major components of a program plan are being implemented in the field. The technique can serve both program accountability and improvement functions. For example, if a program is found to have trouble retaining clients, theory-driven process evaluation can push the inquiry further to find out what is going wrong and impeding retention. Theory-driven process evaluation basically uses program theory, especially its action model portion, as a framework for assessing the implementation process. Four issues are especially pressing when designing and conducting a theory-driven process evaluation. They range from communicating with stakeholders to combining qualitative and quantitative research methods, as explained below.

1. *Briefing Stakeholders on the Purposes, Strategies, and Procedures of Evaluation.* Evaluators need to meet with key stakeholders to discuss the purposes, strategies, and procedures of the upcoming evaluation. The meeting is a good opportunity to obtain stakeholders' support and hear their input.

2. *Clarifying the Stakeholders' Program Theory Concerning Implementation.* Before proceeding with theory-driven process evaluation, stakeholders' program theory, especially as it pertains to the program plan and the action model, must be clearly communicated. (The material presented in Chapters 2 and 3 is a guide for evaluators beginning such a task with stakeholders.) For programs already implemented and essentially matured, agreement among stakeholders about what the program plan should look like comes fairly easily. But even if some components of the program plan do spark disagreement between key stakeholders, that is not an obstacle to evaluation. Rather, disagreement means that evaluators should test various hypotheses while investigating the implementation. Suppose key stakeholders in a program argue about who should be charged with implementing the program—professionals or trained peer volunteers? If implementers currently delivering services come from *both* these groups, evaluation can ask about the relative quality of service delivery by the two. Resulting data would be useful for settling differences among stakeholders as they continue planning future programs.

3. *Research Methods for Theory-Driven Process Evaluation.* Program theory encompasses a variety of elements. In accord, it often requires a combination of quantitative and qualitative methods of data collection.

4. *Freewheeling Data Collection.* The conceptual framework of program theory is an effective guide as evaluators strive to focus on central issues. It is, again, a guide; it should not be regarded as a boundary, barring evaluators from examining important issues outside the framework. In fact, in the course of an investigation of the issues constituting the framework of program theory, important questions beyond its scope tend to be generated (an illustration is included in Chapter 9). Pursuing these questions often provides further enlightenment information for stakeholders' use.

Examples of Theory-Driven Process Evaluation

Comprehensive theory-driven process evaluation is associated with certain strategies and approaches from the taxonomy. Two evaluations are discussed here to show some of the possible functions of this kind of evaluation.

Evaluating a Workplace Smoking Policy. Gottlieb, Lovato, Weinstein, Green, and Eriksen (1992) evaluated the implementation of a restrictive smoking policy for employees of a large, state-run human services agency. The program theory for this workplace policy included four elements: concept, context, process, and outcomes. Its terminology differs, but this model of the workplace smoking policy addresses issues similar to those in the conceptual framework that this book outlines. For example, the category *concept* consists of indicators such as goals/assumptions, nature of the policy and change, and development and support of policy. These mirror to a degree the program rationale; intervention protocol; and, partially, the implementing organization, from the conceptual framework. Gottlieb and colleagues stated explicitly that their model was constructed based upon established literature; it is not clear, though, if stakeholders' input contributed to construction of the model. Gottlieb's team used quantitative methods such as sampling and survey to collect social and demographic information and review employee opinion on several issues: participation in policy development, compliance, and the policy's impact. Qualitative methods such as group interviews, individual interviews, and solicitation of written comments were added to gather information about whether the policy had affected the relationship between smokers and nonsmokers, whether it had made any impact on smoking cessation, and how infractions had been managed. Findings for the four components follow.

- *Concept.* The program rationale was based on needs assessment and was supported by public sentiment. However, a majority of employees said they had had little opportunity to provide input during the formulation of the policy.
- *Context.* The policy affected workers in offices, such as clerical staff, but did not strongly affect employees who spend a major part of the workday in the field. Furthermore, implementing the policy was easier for large work sites than for smaller ones.
- *Implementation Process.* The great majority of respondents supported the policy, but many employees also said they were personally unwilling to report violations. Employees seriously doubted the confidentiality of their complaints. Many supervisors, too, were unwilling to report violations, particularly those at the hands of productive workers. The "designated smoking area" policy had generated some confusion. Issues raised included whether employees could move their work into a break room in order to smoke and whether nonsmokers should be compensated for unused smoking breaks.

- *Outcomes.* In general, nonsmokers perceived that air quality in their work areas had improved. Smokers, on the other hand, perceived that air quality in designated smoking areas had suffered. Nonsmokers' satisfaction with the policy increased over time, whereas smokers' satisfaction decreased.

Findings from the quantitative and qualitative data prompted the authors to suggest practical improvements to the policy and other similar ones. The suggestions were to (a) provide opportunities for employees to join in policy formulation and implementation, (b) provide training for middle managers on how to communicate the policy and enforce it, and (c) ensure that restrictions about smoking would be equal across job categories.

Evaluating an Anti-Drug Abuse Program. One comprehensive, theory-driven process evaluation that closely mirrors this book's conceptual framework of program theory is an evaluation of a large anti-drug abuse program for middle school students in Taiwan (Chen, 1997). The program asked teachers to identify drug-abusing students and provide them with counseling services. A small group of top officials within Taiwan's Ministry of Education had designed the program; under the nation's centralized education system, the Ministry of Education approved appointments and salaries of teachers and administrators. When the program began in January 1991, 3,850 students had been identified as active drug abusers. That number declined sharply, plunging 96%, to 154 students by June 1991.

The program's huge success led to a theory-driven process evaluation being conducted to examine how the program had been implemented. Hopes were that this program's example could foster the smooth implementation of other programs. The anti-drug abuse program featured a documentary program plan, but it was incomplete in comparison to the action model or program plan illustrated in Table 7.1. Acting as facilitators, evaluators convened separate focus group meetings with top officials of the education ministry and with teacher representatives to obtain the information needed to complete the program plan. (The separate meetings acknowledged teachers' tendency to be silent in the presence of top officials, who have much more power than teachers do.) Evaluators played the role of facilitators and consultants, helping these key stakeholders develop their program theory. The final version of the program plan ultimately used for evaluation had been agreed to by both groups; the plan is presented on the left side of Table 7.1.

Table 7.1   The Spring Sun Program: Program Plan Versus Actual
Implementation

Program Components	Program Plan	Actual Implementation
Target population	All drug-abusing student	Only those drug-abusing students who were easy to reach
	Drug use to be verified through urinalysis	Urinalysis collection environment was not controlled
Implementers	Teachers provided with adequate drug abuse treatment training and information	Teachers lacked adequate drug abuse treatment training
Intervention protocol	Primary: High-quality counseling	Counseling mainly involved use of admonishment, threats, and encouragement
	Secondary: Drug education classes	Drug education classes were offered
Service delivery protocol	Compulsory individual counseling	Compulsory individual counseling, but with problems such as lack of plan and objective
Implementing organization	Every school	Smaller schools had difficulty implementing the program
Linking with associate organizations	Effective centralized school system	Communication gap, mistrust between Ministry of Education and the schools
Ecological context Micro	Eliminating video game arcades	Video game arcades still exist
Macro	Strong public support	Strong public support, but problematic education system (elitism)

SOURCE: Adapted from Chen (1997).

The program plan entailed mixing research methods—both quantitative and qualitative—to collect data. For example, quantitative methods were applied to rate teachers' satisfaction with a workshop on drug counseling skills sponsored by the education ministry, whereas qualitative methods were used to probe contextual issues of the teachers' opinions of the workshop. The right side of Table 7.1 displays empirical findings for the program's real-world implementation; comparison of the program theory to the implementation reveals large discrepancies. The program had been carried out, but the quality of services and the system of implementation were far from being impressive. The discrepancies between plan and implementation resulted from a lack of appropriate counseling training, the overburdening of teachers with counseling work with no change to their usual teaching responsibilities, and lack of communication as well as mistrust between an authoritarian ministry and the teachers. The evaluation results created doubt about how a program without strong implementation achieved a 96% decrease in drug abuse in schools.

Theory-Driven Process Evaluation and Unintended Effects

A significant advantage of theory-driven process evaluation is its capacity to detect positive and negative *unintended* effects, if these exist. Its comprehensive examination of the implementation process means that theory-driven process evaluation has a very good chance of revealing important unintended effects of a program. As a matter of fact, both the evaluation of the workplace smoking policy and the evaluation of the anti-drug abuse program discussed above turned up important, though unintended, effects. More on this issue is included in Chapter 10.

PROGRAM MONITORING AND OUTCOME EVALUATION

———•◦•———

In this, Part IV of the book, are found Chapter 8, a discussion of program monitoring, and Chapters 9 and 10, addressing the outcome evaluation strategy used to seek credible evidence of the effect a program makes. Chapter 8 explains both process monitoring and outcome monitoring, although, technically speaking, process monitoring properly belongs to Part III. Because so many of the principles and techniques of process monitoring resemble those of outcome monitoring, it is convenient to discuss both in a single location.

EIGHT

MONITORING THE
PROGRESS OF A PROGRAM

———•◦•———

S takeholders depend on some basic facts about a program's progress as
they continue to communicate with funding agencies and other interested
groups, and for ongoing internal administrative purposes. Preparing to obtain
those basic facts on a continual basis is a task with which evaluators can help;
it is called *program monitoring*. Once a program monitoring routine has been
developed, stakeholders themselves can perform the evaluation into the program's future. Program monitoring may well be the best area for demonstrating the usefulness of empowerment evaluation. This is because in helping to
launch the ongoing program monitoring, evaluators both build the capacity of
the program and its stakeholders to collect evaluative data, and show them
how to interpret and use their data correctly and meaningfully.

WHAT IS PROGRAM MONITORING?

An evaluation that calls for periodic collection of quantitative information
about a program's process and outcomes is called *program monitoring*.
Program monitoring is helpful as a provider of the kind of basic information
to which nearly all programs wish to have access: a set of vital statistics
concerning the program implementation and outcomes. Data from program
monitoring evaluations are often used as indicators of program performance,
but, really, program monitoring does not provide the in-depth information

made available by other evaluation approaches; namely, process evaluation and outcome evaluation. Program monitoring is a useful counterpart to process evaluation and outcome evaluation, but it is not intended to replace them. The differences between monitoring and evaluation will be discussed later in this chapter, which begins by introducing two common types of program monitoring: process monitoring and outcome monitoring.

PROCESS MONITORING

Process monitoring is the periodic collection of implementation information. Information about basic elements of an implementation such as clients' characteristics and service delivery, collected on an ongoing basis, would rapidly become unwieldy, so evaluators need to understand what portion of implementation information will best serve stakeholders' needs. The action model conceptual framework presented in Chapters 2 and 4 can help evaluators and stakeholders make a determination about what information will be collected. In any case, process monitoring should, at a minimum, involve enough information to enable stakeholders and evaluators to know if a program is serving the right individuals and if they are receiving services as intended by the program designers. Process monitoring that gathers at least the following three kinds of information should prove useful.

1. *Sociodemographic Backgrounds of Clients.* An understanding of the variables of clients' backgrounds—their race, ethnicity, age, gender, education, and marital status—helps stakeholders see who is served by a program. The information may also detect any disparity in services.

2. *Kind of Risk Behavior Clients Represent; Severity of Their Need.* It is good for stakeholders to know as much as possible about the degree of risk behavior their targeted population of clients typically represents. For example, HIV prevention programs meet with several high-risk categories of potential clients for whom interventions could be designed: men who have sex with men, intravenous drug users, commercial sex workers, and recent immigrants. In addition, the severity of various clients' needs is sure to differ, calling for various levels of intervention or treatment. The HIV prevention program might find, for example, that clients in the commercial sex workers category rarely practice safe sex.

3. *Number of Intervention Activities Clients Have Completed.* Stakeholders ordinarily like to keep abreast of information about clients' completion of the intervention activities. For interventions requiring multiple sessions, then, information such as the number of sessions completed by each client needs to be collected.

In this book, a strong suggestion is made to evaluators to include process monitoring in the record-keeping system of all programs with which they work. However, even though it is helpful—and technically simpler and financially cheaper than process evaluation—process monitoring should not be expected to take the place of process evaluation. Process monitoring and process evaluation serve unique purposes.

Uses of Process Monitoring Data

Stakeholders such as funding agencies are given to asking the fundamental implementation questions, "How many clients are served?" "Who are these clients?" and "What kind of services are they receiving?" For programs with process monitoring systems in place, answering is easy. Another advantage, due to process monitoring's collection of standardized data at different points in time, is that the quality of implementation then and now can be compared: Was there a point when this program was doing better? Worse? A program that served 300 clients one year but only 200 the next may need some study by its stakeholders because the decline in client numbers could be an indication that something is wrong.

Process Monitoring Versus Process Evaluation

Process monitoring and process evaluation have important differences between them (e.g., fidelity evaluation and theory-driven process evaluation, discussed in Chapter 7). These differences arise in three general areas: scope of data collection, depth of data collected, and data collection across time. The scope of data collection is much broader with process evaluation than with process monitoring. Process evaluation incorporates many aspects of the implementation process, such as the implementing organization, the ecological context, and the implementers. On the other hand, process monitoring seeks only *basic* information about client characteristics (sociodemographic data, relative risk) and client services (services provided, completion status). In addition, the depth of data collected is greater for process evaluation. Typically, there is little depth within process monitoring: Each person receiving a particular service is counted. There is no insight generated into the meaning behind the numbers. If few minority clients were being served by a program, process monitoring would indicate their low numbers but say nothing as to why they were low. By contrast, process evaluation would go further, providing in-depth analysis to obtain the contextual information capable of explaining the near absence of certain groups from the program.

Finally, data collection across time is a characteristic strength of process monitoring but *not* of process evaluation. Process monitoring collects data at various points across time, always with the same instruments used for the initial collection. This allows for comparison of a single program activity at one point in time with that same activity at another point in time. Comparative information of this sort may not be generated by process evaluation unless its design incorporates process monitoring techniques, which can help it overcome such limitation. In other words, process monitoring and process evaluation, used together, compensate for each other's weaknesses; many stakeholders should find this a worthwhile pairing.

OUTCOME MONITORING

Outcome monitoring is the periodic collection of information on outcomes of a program. More specifically, outcome monitoring is the measurement of a program's outcomes at least once before, and at least once after, the offered intervention has been carried out. Thus, it tracks individual clients even after they complete a program. The purpose of outcome monitoring is to acquire data to increase understanding of whether clients are better off for having received services (better off in terms of outcome measures). For example, outcome monitoring of an alcoholism treatment program would require repeated measures of clients' drinking behavior before and after the intervention. Like process monitoring, outcome monitoring uses a standardized instrument to collect outcome data from clients at different points in time. For example, a program to reduce intolerance of the HIV-positive population might develop an instrument that measures respondents' perceptions of and attitudes toward people infected with the virus. As the program conducts outcome monitoring, this standardized instrument is used to measure respondents' levels of intolerance at least twice—once prior to an intervention and once following completion of that intervention. The instrument might also obtain clients' sociodemographic particulars. Typically, outcome monitoring comprises four phases: identification of goals, identification of goal indicators and data sources, determination of needed background information, and pre- and postintervention collection of data.

Identification of Goals

To encourage "buy in," the goals or outcomes to be scrutinized through the monitoring system are usually chosen by a committee of representatives of various stakeholder groups. For example, the Alabama Tobacco Use Prevention and Control Task Force (Alabama, 2000) charged a committee with setting overall goals for its statewide program. The committee established three goals:

- Prevent youths (under age 19) from becoming users of tobacco products.
- Promote treatment of tobacco dependency through promotion of increased access to cessation programs.
- Reduce exposure to secondhand smoke.

Identification of Outcome Measures and Data Sources

In order to monitor goals, indicators and data for measuring them are needed. The antismoking task force found certain outcome measures available in existing data. To assess attainment of the goal of youths' avoidance of tobacco, for example, they used results of the Youth Risk Behavior Survey (a reported percentage of youths who have never smoked tobacco) for comparison purposes. In many instances, however, data relating to outcome measures do not exist. For example, outcome data on participants in smoking cessation programs is not widely available. For such an eventuality, evaluators should be prepared to devise new instruments for measuring outcomes and other data.

Determination of Needed Background Information

A part of most outcome monitoring projects is the collection of sociodemographic data about a program's clients. These data must be in hand in order to compare the success rates of various subgroups within the clientele.

Pre- and Postintervention Collection of Data. The most challenging aspect of outcome monitoring is getting the necessary data at intervals both before and after the intervention. Data for the point in time before intervention are often collected once clients have been recruited to the program; they can be reached and surveyed without much difficulty then. The real work lies in measuring outcomes following intervention. To accurately gauge clients' progress in outcomes, postintervention measures are required some months (and up to a year) after the intervention is completed; this means following up with clients. Following up with clients of programs that serve transient populations—drug abusers, sex workers, the homeless, and other high-risk groups—is especially difficult. Because members of the evaluation team must track and interview the program participants, the cost of the outcome monitoring contract is increased. (Tracking can be made somewhat easier by asking clients, at the time they enroll, to give addresses and telephone numbers of their relatives or friends, as well as their own—a popular strategy.) Thus, outcome monitoring is both more difficult and more expensive than process monitoring.

Outcome Monitoring Versus Outcome Evaluation

Stakeholders have a great interest in understanding the outcomes of a program. Outcome monitoring and outcome evaluation each provide good information, but in different contexts. Therefore, evaluators need to understand clearly the nature, functions, and limitations of the two. They differ principally in the areas of methodology and interpretation of results. When it comes to methodology, outcome evaluation is more stringent, requiring use of a rigorous design (quantitative, qualitative, or mixed). Often, the outcome evaluation design entails manipulation of intervention conditions, such as creating a control or comparison group. In contrast, outcome monitoring consists primarily of conducting pre- and postintervention measures for a single group and does not require a rigorous design. An exception is the outcome monitoring study that consists of several outcome measures before and after the intervention. In such a case, outcome monitoring can be upgraded to a quasi-experimental design called an "interrupted time series design" (Shadish, Cook, & Campbell, 2002).

Outcome monitoring also differs from outcome evaluation where interpretation of the analyzed data is concerned. Through outcome monitoring, an evaluation can say whether clients do the same, better, or worse following intervention. But by itself, it often cannot produce credible evidence definitively attributing to the intervention those outcomes have resulted; too many potentially confounding factors exist, such as program maturation and historical events (Shadish et al., 2002). Another way of putting it is that simply knowing, from the monitoring data, that postintervention measures are "better" than preintervention measures is not enough to prove that the intervention exerted a specific effect. Because outcome evaluation data come from a rigorous design that is able to manage the potentially confounding factors, results of outcome evaluation are considered defensible evidence of a program's effectiveness or failure.

Strengths and Limitations of Outcome Monitoring

Stakeholders also have a great concern for how clients fare within a program. Although outcome monitoring is not meant for scientific evidence of program effectiveness, one should not underestimate its usefulness to stakeholders during program development. One strength of outcome monitoring is that it provides data that respond, in a timely way, to stakeholders' concerns about clients' progress or lack of progress. Data of this nature will suffice when stakeholders want to strengthen a program, not sum up its effectiveness. Thus, if, according to outcomes, clients are not faring well following intervention, enough information exists to suggest that the program is not working and that modifications that can strengthen the program should be identified. (The types of process evaluation discussed in the preceding two chapters are very helpful for identifying problems with programs.) Most often, no rigorous, time-consuming outcome evaluation needs to verify a program's ineffectiveness. On the other hand, when outcome monitoring data show clients faring better in postintervention outcomes, it can be concluded that the program is promising. The program may, in fact, have made some contribution to clients' progress, but an outcome evaluation will be necessary to answer categorically the question of program effectiveness.

Another strength of outcome monitoring that should not be overlooked is its affordability. Outcome evaluation is usually high-priced and difficult to do. It is unreasonable to expect small, community-based organizations to treat their programs to outcome evaluations; the cost of conducting a randomized experiment can, in fact, equal some organizations' budgets for providing the services they provide. In a case of this kind, funding agencies should be prepared to allow small organizations to rely on outcome monitoring rather than outcome evaluation. Furthermore, outcome monitoring functions as an evaluation capacity-building experience for organizations—a worthwhile investment. Indeed, outcome monitoring is a foundation for outcome evaluation. An organization that masters outcome monitoring is demonstrating a capacity to track client performance and use the data for program improvement. Once an agency can complete outcome monitoring, it becomes easier for it to appreciate and to attempt outcome evaluation. At this level, there will likely be less resistance by program staff to an upgrade to outcome evaluation. Asking an organization to jump immediately into outcome evaluation, with all the associated disruptions, may be counterproductive if program managers and implementers doubt whether the trouble they must take to support evaluation will pay off, given that the resulting data may strike them as not entirely useful to their agendas.

To summarize, a general rule for outcome evaluation of intervention programs is this: When stakeholders need credible evidence of their program's effectiveness, choose outcome evaluation; when they want outcome data to show progress, to communicate with people interested in the program, or for internal management tasks, choose outcome monitoring. Outcome monitoring holds special relevance for organizations with tight budgets. In the same vein, a funding agency making a small grant to a community-based organization can reasonably request the agency to perform outcome monitoring, but expecting an outcome evaluation from it is unreasonable.

Integrating Process Monitoring and Outcome Monitoring in a Program Monitoring System

Process monitoring is the gathering of data concerning the services received by clients from a program, whereas outcome monitoring is the gathering of data concerning these same clients' outcomes following intervention. It makes sense to integrate these two in a *program monitoring system*. Process monitoring and some outcome monitoring can be conducted simply with paper and pencil. An integrated program monitoring system will not be so uncomplicated, usually requiring an electronic information system to store and manage data and build the organization's capacity to implement monitoring. This is the topic of the rest of this chapter.

PROGRAM MONITORING
SYSTEMS WITHIN ORGANIZATIONS

The Rise of the Program Monitoring System

With Congress's passage of the Government Performance and Results Act (GPRA) in 1993, there began a nationwide trend among federal, state, and local agencies to affirm more strongly their accountability for their programs. One impact of the GPRA is that more and more large agencies are instituting program monitoring, and even evaluation systems, to collect the data that assist them in meeting the requirements of the act. This development has also had a trickle-down effect among local agencies and community-based organizations. Program monitoring systems of the kind being so widely adopted are *institutionalized* systems for gathering information, on an ongoing basis, about programs and the activities they sponsor. In other words, the collection of monitoring data should become a routine task in an organization that implements a program monitoring system. Monitoring data usually include implementation and outcome data and can extend to the program planning process. For example, in order to ensure that an organization's program planning process is participatory, a program monitoring system can observe whether, at the local level, community representatives and experts are being invited to join in planning.

Program Monitoring System Elements

A program monitoring system typically integrates four elements: monitoring guidance, an electronic data system, capacity building/technical assistance, and support from top management. Monitoring guidance refers to the process of outlining which standardized data elements will be collected, how they will be collected, how they will be reported to a central site(s), and how the information they generate will be utilized. The electronic data system is required for most program monitoring systems to enable them to process data. Given the current availability of information technology, it is feasible to build a web-based computerized system that electronically links all of an organization's local units to a central information center. Within such a system, data can be easily transmitted, stored, managed, and analyzed—at least when appropriate data managers are engaged.

This brings us to the element called capacity building/technical assistance. Maintaining a program monitoring system depends on local units' and implementers' mastery of the forms and other requirements mandated by whatever monitoring guidance is given. The local units and implementing staffs must integrate these requirements into their day-to-day activities. However, not every local unit and staff is immediately capable of doing so, and the first question asked must be whether the local organization has sufficient manpower for collaboration. A program monitoring system requires designated staff responsible for caring for the system. It is not unusual for funds to be provided to local units that lack a staff to manage data. A second question as to capacity concerns whether the local staff has the skills and knowledge needed to collaborate. The quality of monitoring data stems from a staff's capacity and willingness to daily and correctly record their activities.

Most often, staff will need intensive training. Furthermore, because the first steps of launching a program monitoring system within local facilities tend to be dogged with various difficulties, the larger organization should be ready with generous technical assistance for local staff. Such generosity is more forthcoming when creation of the program monitoring system has the support of top management. Developing and launching a program monitoring system—notably one at the state or national level—is a complicated process. It is likely to demand certain changes to the structure and activities of units across the entire organization. Predictably, such a top-to-bottom project will encounter problems, complaints, and resistance to change. With strong support from top management, these can be overcome; without it, the monitoring system seems almost to fall apart after any stumble, and it can easily end up terminated. Unless the program monitoring system is backed unquestionably by those in highest authority, an attempt to build it simply cannot be recommended.

Developing a Program Monitoring System

Either a top-down approach or a participatory approach can be adopted for the development of a program monitoring system. Under the top-down approach, a small group of experts and top managers devise the system and ask employees down the ladder to implement it. The high efficiency of this approach is one of its advantages. The experts need no more than several months to work out the instruments, data collection and reporting procedures, and staff training in order to be ready for implementation. The top-down approach is at a disadvantage, however, in that implementers and administrators at the local level may not understand, support, and commit to launching the system (Chen, 1997). In contrast, the participatory approach consists of extending an invitation to representatives of various stakeholder groups to help develop the program monitoring system. Although the group nature of the endeavor entails more time and coordination of effort, when a system has been created in this way implementers and other stakeholders are more likely to accept and support it. In addition, the presence of stakeholders' input tends to help the system better respond to local-level needs and interests, making it more useful to various organization units. The system's success in capacity building/technical support also seems to burgeon under the participatory approach. The participatory approach to creating a program monitoring system is highly recommended; an example of this approach in action follows here.

Developing and Implementing a
Program Monitoring System: An Example

Starting in 1998, the Centers for Disease Control and Prevention (CDC) began to develop a program of monitoring and evaluating HIV prevention efforts (CDC, 2001; Chen, 2001). Its program monitoring system formed the base of the system overall; the central experiences and issues pertaining to the new program monitoring system are explored below. The CDC's development and implementation of a comprehensive program monitoring system for anti-HIV efforts was expected to affect not just CDC internal operations but also the activities of 65 state or territorial health departments and several thousand community organizations across the nation. With a change of this magnitude, organizational or even institutional change might be demanded to finalize the design and implementation of the system.

Potential barriers to the accomplishment of change had to be identified and removed. Consultation with key informants across the various stakeholder groups was a means of discovering and understanding some of these barriers, which included overburdening of workers, suspicion of CDC motives, fear of being ignored as a voice in the implementation effort, and lack of capacity to fulfill new requirements. To be specific, one barrier to change was that the new program monitoring system would increase the workload of health departments and community-based organizations because they had to either modify existing data systems or develop new systems and collect more data. Another barrier was the health departments' suspicion that the CDC might want these monitoring results to use as punitive criteria when awarding grants. A third barrier to change was that health departments and other stakeholders feared that their input to CDC would be limited, the system would be created essentially without any consultation with them, and its implementation would be mandated. Finally, many stakeholders to be affected by the program monitoring wondered if they had enough money, manpower, and know-how to respond to the requirements coming from the CDC.

These barriers were not beyond addressing. To do so, the CDC convened several meetings with representatives from health departments, community-based organizations, and other stakeholder groups, asking for comment on issues such as evaluation purposes and content, relevant data elements, methodologies, and technical assistance. The meetings were also a vehicle for discussing the need for a program monitoring system, the expected uses for the data, and the kind of capacity building/technical assistance really available. The stakeholders steadily added their input as the monitoring guidance evolved, as the electronic data system was outlined, and as components of capacity building/technical assistance took shape (e.g., training of health department staff and CBO representatives, development of peer review manuals). In addition, new funding was designated for evaluation activities, enhancing health departments' capacity to make a commitment to their performance.

Implementation of a Program Monitoring System

In making the program monitoring system happen, the CDC and the stakeholders had to consider issues of program planning, implementation, and effectiveness. Documentary monitoring guidance (direction about the collection of particular data) was completed and implemented in 2001. According to this monitoring guidance, information collected for the monitoring system would comprise the following: assessment of HIV prevention community planning, design and evaluation of intervention plans, monitoring and evaluation of implementations of HIV prevention programs, evaluation of linkages among the prevention plans and resource allocation, and evaluation of outcomes of HIV prevention programs. Following suggestions made by health departments, phasing in these program activities smoothed their implementation. Initially, activities involved in the design and evaluation of intervention plans, and in the evaluation of linkages between plans and resource allocation, were implemented. Next came the monitoring and evaluation of implementations of anti-HIV programs. Later, evaluation of program outcomes was undertaken.

Lessons Learned From This Program Monitoring System

The participatory approach to developing the program monitoring system at the CDC ensured that a majority of stakeholders supported such evaluation and cooperated throughout the development process. It is important to note, though, that stakeholder participation remained a give-and-take exchange. It is the rare evaluator who is given everything he or she might want, and, in this experience with the CDC, evaluators compromised on both the scope and the focus of the monitoring system. They negotiated the portions of the evaluation to be scheduled sooner and those scheduled later, the requirements that would be revised to better reflect reality, the assistance to be extended to health departments, and the ultimate uses of evaluation data. A program monitoring or evaluation system in place and functioning adequately can make a substantial impact on a program's quality. And indeed, early comment on the new CDC system from state health departments indicated that the range of program monitoring data now available had been useful in their program planning and implementation efforts.

NINE

EVALUATING OUTCOMES

*Efficacy Evaluation
Versus Effectiveness Evaluation*

———◆•◆———

The resources committed to and used by a program are many. Therefore, when a program of any size reaches maturity, there are likely to be funding agencies, decision makers, program managers, implementers, and citizens who have a serious interest in understanding the ultimate effect of the program. Outcome evaluation serves this purpose. Outcome evaluation is a rigorous assessment of *what happened* thanks to a program. Three general types of outcome evaluation are presented in this book, and this chapter focuses on two of them: efficacy evaluation and effectiveness evaluation. Both of these aim to answer the question, "Is this program achieving its goals?" To go beyond merit assessment and establish reasons why a program achieves or does not achieve its goals, *theory-driven outcome evaluation,* the topic of Chapter 10, is the appropriate tool.

GUIDELINES FOR CONDUCTING OUTCOME EVALUATION

1. Planning the Outcome Evaluation

Outcome evaluation is much more expensive and complicated than other types of evaluation, but, because it supplies such important information, funding agencies often require its use. Because of the cost and effort, outcome evaluations should be carefully planned to the last detail. Often, instead of rushing ahead with outcome evaluation directly, evaluators would be wise to conduct a pre-assessment of the mature program, called *evaluability assessment,* the topic of the next section. It can determine if the program is truly ready for rigorous outcome evaluation, and it, too, requires thoughtful planning in advance. Although outcome evaluation can be applied to start-up and established programs alike, in general, the process is easier with start-up programs. Working with start-up programs, evaluators have an opportunity to guide stakeholders' structuring of their program, introducing rigorous design from the start. They also have the advantage of selecting from the variety of evaluation strategies and approaches this book has elaborated, those that will ensure good-quality program planning and implementation. Long-running, established programs that bring in evaluators after considerable periods of implementation lack these design advantages. However, this does not mean outcome evaluation cannot be carried out in established programs. In such programs, evaluators need to creatively put together an evaluation design that takes into consideration factors such as the availability of data kept by the implementing organizations, the patterns of client enrollment, the availability of relevant surveillance data, and the availability of comparison groups.

While planning an outcome evaluation, a crucial decision for stakeholders and evaluators is whether it should be an efficacy evaluation (evaluation within ideal circumstances) or an effectiveness evaluation (evaluation in the real world). Established programs, operating in the real world as they do, have little choice: only effectiveness evaluation suits. Start-up programs, in contrast, have a choice to make, upon which hinge the requisite evaluation design, the results of evaluation, and the eventual utilization of those results. Intensive discussion of the factors in this decision comprise much of the information in this chapter.

2. Applying an Appropriate Conceptual Framework

Efficacy and effectiveness evaluation reflect disparate assumptions, and each set of assumptions generates a distinct framework for evaluation. A common problem in evaluation involves a mix-up of these frameworks: The evaluator evaluates strictly in accord with principles and strategies from an efficacy framework, when what stakeholders really had in mind was effectiveness evaluation. This chapter clearly delineates each framework, providing an understanding that should help the evaluator avoid such mix-ups.

3. Selecting a Research Design

With its burden of producing credible evidence of a program's effect, outcome evaluation must employ rigorous design. Efficacy evaluation follows closely the traditional scientific paradigm, very often its reliance on randomized, controlled experiments. In its real-world setting, effectiveness evaluation, by contrast, may have difficulty with or be undesirable for randomized experiments. Evaluators often need to consider and use other research designs, such as quasi-experimental designs. Whatever the choice of research design, its integrity and that of the associated research procedures must be maintained throughout the evaluation.

4. Data Collection

The very least that efficacy and effectiveness evaluations must accomplish is collection of data about intervention and outcome. Because an efficacy or effectiveness evaluation should be as sensitive as possible to detectable changes in outcomes, precise instruments are often needed for measurements of outcomes "before and after" an intervention. It can also be useful for obtaining data, such as clients' and implementers' characteristics, because that information can contribute to the fine-tuning of the assessment and direct the interpretation of results. Data on the fidelity of implementing intervention are also needed to analyze the intervention effect.

EVALUABILITY ASSESSMENT, UNDER
THE DEVELOPMENT FACILITATION STRATEGY

Not every established program is immediately suitable for outcome evaluation. *Evaluability assessment* can be used to determine whether a rigorous outcome evaluation is warranted (Smith, 1989; Wholey, 1987, 1994). If the assessment shows the program to be evaluable, outcome evaluation is justifiable. If the program proves *not* to be evaluable, evaluators work with stakeholders to foster its evaluability. Thus, evaluability assessment has a strong flavor of the development feature and, for this reason, rests within them development facilitation strategy in the taxonomy (see Table 3.1). In fact, it is very often a funding agency or program manager who, recognizing that a program needs modification, initiates evaluability assessment. Thus, a major reason for conducting evaluability assessment is to assist various stakeholders with improving their program.

Techniques of evaluability assessment include site visits, records review, intensive interviews, and focus group meetings with key stakeholders, all looking for a program's manifestation of the following four criteria (Wholey, 1994):

- The goals, objectives, important side effects, and priority information needs of the program are well defined.
- Goals and objectives of the program are plausible.
- Relevant performance data can be obtained.
- Intended users of evaluation results agree on how they will use them.

Without a set of clear, consistent goals, along with evidence that a program is being implemented in the direction indicated by those goals, attempting to carry out outcome evaluation is not prudent.

For example, deriving the needed outcome measurements may not be possible with a program lacking clear and consistent goals. Proceeding with outcome evaluation via an arbitrary choice of such measures—when real goals, are, in so many words, unknown—may well find the evaluator in a scenario in which stakeholders challenge the relevance and significance of the evaluation. They doubt that the proper goals were evaluated, or feel that the evaluator's work fails to reflect their own work. In short, if a program has been implemented *without* clear goals, there is little reason to expect it to have attained them. To conduct outcome evaluation of such a program is a waste of time and resources.

One goal that *must* be clearly identified by any program is the reaching of a particular, well-defined target population. Most HIV prevention programs, for example, plan to reach out to one or more high-risk groups: teenage runaways, undocumented immigrants, and others. To reach its intended clients, a program needs effective strategies. These must be spelled out, and records must demonstrate plainly who the program has already reached and exactly what services they received in order for outcome evaluation to be meaningful. The program without an ample supply of this information needs to be evaluated, but not in terms of its outcomes. Rather, its structure, priorities, and resource allocation must be evaluated before its effectiveness can be. In other words, a poorly implemented program will benefit more from evaluations such as needs assessment or process evaluation. A rule of thumb is that a program unsuited to be evaluated in terms of its effectiveness probably needs to be redeveloped and restructured.

NONLINEAR TRANSITIONS BETWEEN PROGRAM STAGES

The foregoing discussion of evaluability assessment shows, between the lines, that programs' progress through developmental stages is *non*linear. Just because a program has operated for years in the mature implementation stage does not necessarily mean that it has moved on to the outcome stage. But it can always be readied for that move to outcome. Evaluability assessment's real purpose is helping stakeholders redesign programs to make them ready for outcome evaluation. To do this, the programs must return to the planning stage first, then jump back to the outcome stage once evaluability has been created. Importantly, evaluability assessment is just one option available to evaluators seeking to assist stakeholders with such redesign. It is probably the best option for the evaluator trying to help build consensus on program goals and objectives, a prerequisite for successful design and use of outcome evaluation. At times, however, stakeholders ask for more than enhanced evaluability of their program; the development facilitation strategies explored in Chapters 4 and 5 will provide other options for them. Those who worry about the soundness and plausibility of the action and change models underlying their program, for instance, can make good use of development facilitation evaluation during program redesign.

EFFICACY EVALUATION OR EFFECTIVENESS EVALUATION?

When stakeholders are interested in knowing how well a program performed in terms of goal attainment (performance assessment strategy in Table 3.1), two kinds of outcome evaluation can be used: *efficacy evaluation* and *effectiveness evaluation.* Efficacy evaluation involves assessing the effect that an intervention has when conditions are ideal.

According to Flay (1986), efficacy trial or evaluation is characterized by strong control in that a standardized program is implemented in an uniform way to a specific and narrowly defined homogeneous target group. Due to the strict control and standardization, the evaluation can provide highly convincing evidence of the effect of an intervention. Effectiveness trial or evaluation assesses the effect of an intervention in real-world conditions. According to Flay (1986), effectiveness evaluation is characterized as the standardization of availability and access among a broadly defined population while allowing implementation and level of participation to vary on the basis of real-world settings.

Efficacy evaluation is highly attractive to scientists because of its scientific merits. It closely follows scientific principles to conduct assessment, with evaluators or researchers regulating factors such as research conditions and who participates in the program. The aim is to manipulate conditions to ensure that a precisely measured "dosage" of the intervention is delivered, in standardized fashion, to clients. Tight research controls ensure the continued integrity of intervention conditions and study design throughout the evaluation. This is largely why evidence from efficacy evaluation is usually very precise and highly defensible, at least from the traditional scientific perspective.

Conducting an effectiveness evaluation is quite challenging. The real world is the matrix for effectiveness evaluation of interventions. Generally speaking, it requires greater consumption of time and resources than does efficacy evaluation. Furthermore, because effectiveness evaluation operates in the real world, stakeholders' support and collaboration can be particularly vital to it. As an effectiveness evaluation proceeds, a program's usual implementers provide service to typical clients, who may not be, respectively, very well trained or minimally cooperative. In efficacy evaluation, evaluators have the option to work with implementers of the utmost competence, as well as with especially motivated clients; but effectiveness evaluation demands the program's common climate. For example, during real-world research, data frequently are, of necessity, collected by implementers, so one issue that matters deeply in effectiveness evaluation is whether particular implementers are or are not capable of and committed to collecting the data. Compared to efficacy evaluation, then, the research procedures that constitute effectiveness evaluation are less clear-cut, more messy—as are the results of effectiveness evaluation. Another important feature of effectiveness evaluation is its clear object:

evaluation results with implications useful for the betterment of practices or programs, either existing ones or future ones.

More is said in this chapter about effectiveness evaluation than efficacy evaluation because the methodology of efficacy evaluation and its applications are well covered in literature. Many fewer discussions in literature cover effectiveness evaluation.

Choosing Efficacy or Effectiveness Evaluation

The first matter to resolve, when stakeholders propose evaluation to assess a program's attainment of goals, is whether an efficacy evaluation or an effectiveness evaluation should be used. Design of the outcome evaluation, as well as choice of methodology, differs for the two approaches (as will be explained later in the chapter). There are several guidelines for identifying the appropriate outcome evaluation approach.

Conditions Favoring Use of Efficacy Evaluation

- Stakeholders such as funding agencies and program designers seek a demonstration project maximizing opportunity of an innovative intervention's ability to produce desirable effects under ideal conditions.
- Evaluation aims include developing scientific theory and knowledge.
- The evaluation is unburdened by a need for information for stakeholders' immediate practical use.

Conditions Favoring Use of Effectiveness Evaluation

- Stakeholders question the feasibility of implementing a new program in the real world.
- Stakeholders are curious about the effects of a new program, or of an ongoing program, in the real world.
- Stakeholders require the evaluation to be relevant and of practical benefit to their practices related to a program.

EFFICACY EVALUATION'S CONCEPTUAL FRAMEWORK AND METHODOLOGY

Efficacy evaluation bears out the assumption that only under tight research controls can the actual intervention effect be assessed validly. By adhering to

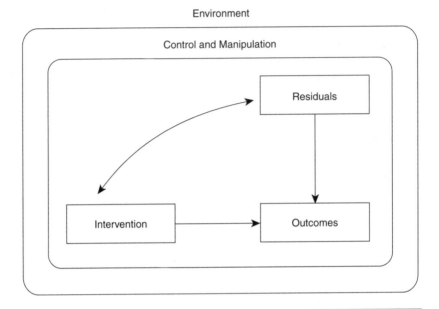

Figure 9.1   Conceptual Framework of Efficacy Evaluation

the traditional scientific research framework, efficacy evaluation has come to be regarded by many scientists as the standard for *scientific knowledge acquisition* concerning intervention in biomedical research. The view has also been adopted in health promotion and social intervention. Efficacy evaluation focuses on issues related to internal validity. It asks how to rule out rival hypotheses and confounding factors capable of impeding assessment of an intervention effect. Any practical implications that the findings of efficacy evaluation might hold are the topics or tasks of future inquiries.

The outline of the conceptual framework of efficacy evaluation is modeled in Figure 9.1.

Figure 9.1 indicates that the main task for efficacy evaluation is to assess, precisely and without bias, the effects of the intervention on the outcomes. However, the assessment is likely to be interrupted with three things: environmental intrusions or disruptions, uncertainty of intervention conditions, and the possible existence of rival hypotheses.

Efficacy evaluation that closely follows scientific procedures must deal with these three problems:

1. *Control the setting.* Efficacy evaluation usually controls or manipulates the settings to seal off or reduce the environmental influence or intrusion. The ideal is a laboratory kind of setting.

2. *Control the intervention conditions, clients, and implementers.* This is necessary to reduce the number of other potential factors that may affect the outcomes at the same time. Evaluators and implementers deliberatively recruit homogeneous and motivated clients to participate in the trial to minimize the potential effects of variation among clients' characteristics. They use highly trained implementers and highly standardized intervention procedures to minimize the size of residuals.

3. *Assign clients randomly into the intervention conditions.* Another major problem in outcome evaluation is that some of the residuals (which are known as *rival hypotheses* or *confounding factors*) may correlate with the intervention. The recruitment of a homogeneous group is helpful but does not eliminate these correlations. For example, in a clinical trial, a group of patients may be homogeneous in terms of age and sickness. However, the patients may still be different in terms of history of sickness, gender, weight, and so on, which may correlate with the intervention. Efficacy evaluation deals with this problem by randomly assigning clients to the experimental or control groups. Randomization generates equivalent experimental and control groups for an unbiased assessment of intervention effect.

The combination of these three procedures is often called *randomized controlled experiments.* This randomized controlled design has been regarded as the most rigorous for efficacy evaluation. Efficacy evaluation may use quasi-experiments (Shadish et al., 2002), which are acceptable but do not provide the strongest evidence. Preexperimental designs, such as surveys and correlation studies, are not desirable in efficacy evaluation.

Efficacy evaluation is a highly popular approach for conducting outcome evaluation in many disciplines. In health promotion, for example, the bulk of outcome evaluations performed are efficacy evaluations (Clark, 1995; Glasgow, Lichtenstein, & Marcus, 2003; Weisz, Weiss, & Donenberg, 1992). The popularity of the approach has many reasons:

- Panels that review grant proposals for the National Institutes of Health and many other federal funding agencies favor efficacy evaluation. It exhibits the scientific rigor such reviewers prefer.

- Reputable journals prefer to publish articles describing well-controlled efficacy studies.
- By creating ideal intervention conditions, researchers can control participants and other matters, making evaluation and research much more manageable.
- The belief is common (and strongly held) among scientists that an intervention's efficacy needs to be evident *before* any examination of its effectiveness or generalizability can have meaning.

Research Example of Efficacy Evaluation

There is much written about efficacy evaluation in the literature, much of it recent. However, for two reasons, a choice was made to illustrate the discussion here with a relatively old study of treatments for public-speaking disorder (Paul, 1966). First, Paul's study makes plain the characteristics, focus, and strategies of efficacy evaluation. Second, an intervention similar to the one evaluated by Paul has also been assessed by other researchers using effectiveness evaluation. The latter case will be presented later in this chapter, allowing comparison and perhaps providing greater insight into issues of efficacy versus effectiveness evaluation. Paul's (1966) study compared the efficacy of psychotherapy and behavior therapy for helping individuals with public-speaking disorder. Paul saw that a tightly controlled experiment would be needed in light of potential intrusions by various extraneous factors, including patients' characteristics, therapists' characteristics, and therapeutic circumstances. Of 710 undergraduates enrolled in a public-speaking class, Paul selected a highly homogeneous group of 97 students who had scored high on performance anxiety scales, writing that he surmised the selected subjects were "good bets" for his research purpose. The 97 subjects were randomly assigned to one of five groups: psychotherapy, behavior therapy, attention placebo, no treatment (this group was interviewed and promised treatment later), and no contact (a group never contacted for the project at all). Each kind of treatment was delivered by one of the area's most highly regarded therapists. Because the therapists' personal characteristics presented extraneous factors, each therapist was required to use all five treatment modes with different participants. Paul's study results showed that the groups assigned psychotherapy, behavior therapy, and attention placebo made significantly more progress than those in the no-treatment and no-contact groups.

Furthermore, the behavior therapy group had seen stronger improvement than the psychotherapy group. Paul's study offers—through use of experimental design and controls, a homogeneous groups of subjects, and top-notch therapists—some clear-cut evidence that behavior therapy surpasses psychotherapy in treating public-speaking disorder. Paul's work was a genuine, and useful, contribution to the advancement of knowledge and the development of the theory of therapy.

EFFECTIVENESS EVALUATION
IN THE CURRENT STATE OF THE ART

Effectiveness evaluation scopes out the effect of an intervention in its real-world milieu. Its end is to provide practical information about ways to improve current and future programs. Principles and strategies of effectiveness evaluation differ greatly from those of efficacy evaluation. Operating in the real, uncontrolled (scientifically speaking) world, effectiveness evaluation is bound to confront many difficulties not addressed within the framework of efficacy evaluation. Thus, it cannot take on identical procedures. Unlike the controlled setting, the real world can be a messy place to conduct outcome evaluation: a setting in which the delivery of services and the gamut of clients are as natural as daily practice in the field. In effectiveness evaluation, the clients studied are not specially screened and recruited, as they are in efficacy evaluation. Similarly, the observed implementers are not restricted to those most fully trained, committed, and enthusiastic. The implementing organization is practice-based, not a research institute. And in effectiveness evaluation, evaluators have no direct control of implementation processes, so environmental intrusions are very likely to occur.

Effectiveness evaluation has a long history of use with established programs in areas such as education (e.g., Head Start), criminal justice (e.g., DARE), and welfare reform programs (Rossi et al., 2004); it is not popular for evaluating innovative interventions in health programs. For example, in the literature of health promotion and other behavioral interventions, very few effectiveness evaluations have been reported in journals as compared to the number of published efficacy evaluations (Clark, 1995; Glasgow et al., 2003; Weisz et al., 1992). The following factors contribute to the relative unpopularity of effectiveness evaluation:

- Its research conditions typically discourage a rigorous design that meets the scientific standard, although it nevertheless demands much more time and effort than efficacy evaluation. Academe's current rewards structure holds effectiveness evaluation in disfavor, meaning that a proposed effectiveness evaluation is less likely than efficacy evaluation to be awarded grants.
- It is widely held that efficacy evaluation must precede effectiveness evaluation. Too often, however, the effectiveness evaluation never follows. It is common for researchers or evaluators to move instead to another efficacy evaluation. Researchers or evaluators strongly committed to an efficacy-to-effectiveness cycle are uncommon.

Concern over the disparity between the amount of literature on efficacy evaluation and that on effectiveness (Oldenburg, Ffrench, & Sallis, 2000; Green, 2001; Glasgow et al., 2003) is growing. This momentum may well be reinforced by the budding interest in "translation research"—translating research into evaluation practice—which would necessitate increased publication of effectiveness research. To stimulate this correction, the field needs a better conceptualization of and a useable framework for effectiveness evaluation in action. This need provides the topic of the rest of the chapter.

RECONSIDERING THE EFFICACY-TO-EFFECTIVENESS PARADIGM

As indicated above, scientists tend to believe that efficacy evaluation needs to precede effectiveness evaluation (Flay, 1986; Greenwald & Cullen, 1985), leaving effectiveness evaluation to play second fiddle. They maintain that only when efficacy evaluation identifies desirable outcomes for an intervention should effectiveness evaluation be brought into play, seeking the same effect in a real-world surrounding. This paradigm, rooted in biomedical research norms, has certain ethical and financial merits. For instance, in biomedicine, unproven drugs and treatments ethically cannot be used with ordinary patients. Conducting efficacy evaluation first can establish any potential benefits of new medical tools, justifying human trials. In addition, efficacy evaluation in biomedicine is much less costly than effectiveness evaluation and consumes less time. This makes it quite feasible to consider completing a number of efficacy trials in advance of performing an effectiveness trial. Furthermore, there is an important difference in biomedicine between medical procedures and other types of interventions: the attainability of a kind of closure for treatment assessment. Chemical compounds and definitive courses of treatment maintain their integrity whatever the setting of the service delivery (laboratory or real world). Thus, biomedical effects established through efficacy evaluation hold great promise of resurfacing when effectiveness evaluation is conducted.

Indeed, the efficacy-to-evaluation paradigm works very well with the needs of biomedical research. But clearly, those may not be the needs of a developing prevention or health promotion program, nor of social programs. First of all, biomedicine's need for solid ethical grounds justifying the implementation of programs is much less pressing for these other kinds of programs. This is because, in the latter, the risk of severe injury or death resulting from an intervention is so tiny. Health promotion and social interventions comprise education, counseling, skill building, role playing, group discussions, outreach, and social marketing rather than toxic chemicals and somatic invasion. At worst, health promotion and social interventions may make no difference; they would not, in all likelihood, do any severe physical or psychological harm. Furthermore, effectiveness evaluation has to go through the same institutional review board process as efficacy evaluation. Human subject protection is applicable to both in health promotion and social programs. In truth, many (perhaps most) interventions transpire at the hands of ordinary organizations, without evaluator or researcher assistance. On an ongoing basis, decision makers in both private and public sectors launch all kinds of programs meant to foster safety, productivity, social harmony, and well-being.

In sum, biomedical research's efficacy-to-effectiveness paradigm on ethical grounds does not apply as strongly with social and health promotion intervention. A second difference is that the financial savings that biomedical research incurs by using efficacy evaluation dries up where social and health promotion programs are concerned. Whereas several dozen patients may suffice for a laboratory-based biomedical efficacy evaluation, efficacy evaluation with social and health promotion programs involves large samples in order to validate statistical power. The sample size in efficacy evaluation may be less than in effectiveness evaluation, but not by much. Furthermore, establishing and controlling conditions and research procedures entails generous spending of time and resources, to the point that it is unclear whether efficacy evaluation is more expensive than effectiveness evaluation.

Another difference is that in biomedical research, because of the relatively low cost and short turn time, efficacy evaluation of a drug or treatment can be carried out many times before an effectiveness evaluation, which further strengthens the efficacy-to-effectiveness paradigm. This is not applicable to social and health promotion programs because the higher costs and longer turn time mean that evaluation can be done only once. Stakeholders and evaluators need to make a choice: efficacy or effectiveness.

Finally, returning to the idea of the stand-alone nature of the drug or medical therapy in biomedical research, drug and medical therapy in efficacy evaluation and effectiveness evaluation are the same regardless of different settings. This does not hold true in social, health promotion, or prevention types of programs. Human behavior is more volatile than human physiology because behaviors vary with shifting context: the social milieu. It follows that clients' behavior in controlled situations can differ markedly from their behavior in social situations that have not been manipulated. The artificiality of results obtained in controlled situations is repeatedly documented in social research (e.g., the famous Hawthorne studies discussed in Rossi et al., 2004). In other words, lessons learned about an intervention via efficacy evaluation often will not be the same as lessons taught by effectiveness evaluation. The degree of confidence evaluators should have about generalizing efficacy findings to effectiveness in the real world is a great unknown. Often, results of efficacy evaluation prove too optimistic for the realities of effectiveness research. In short, the grounds for the efficacy-to-evaluation paradigm weaken dramatically when evaluation steps away from biomedicine into the area of health promotion and social concerns.

Despite these things, abandoning efficacy evaluation altogether would be extreme. It plays a vital role in the development of scientific knowledge, generating information that can be quite useful when effectiveness evaluation is being designed. The discussion above does imply strongly, however, the need to appreciate and fund effectiveness evaluation on its own merits. Prior efficacy evaluation need not figure in the equation outside of biomedicine. Effectiveness evaluation develops practical knowledge, or knowledge of practice. Health promotion and social programs practice in the real world. Thus, funding agencies really should not place so much emphasis on efficacy over effectiveness evaluation. The relationship of the two is out of balance, which may hinder the future advancement of our knowledge and skills for actually solving social problems or promoting health.

This book proposes a bilateral efficacy and effectiveness paradigm for social and health promotion programs. In this paradigm, efficacy evaluation is used to test intervention derived from scientific theory that helps to develop universal proposition and laws. Effectiveness evaluation is used to test interventions that are interesting to stakeholders and relevant to their practice. The idea and design of intervention of effectiveness trials can benefit from similar or related efficacy evaluations. Similarly, the idea and design of intervention of efficacy trials can also benefit from the effectiveness evaluation. The funding agencies should fund effectiveness trials as frequently as efficacy trials.

EFFECTIVENESS EVALUATION ADOPTS EFFICACY EVALUATION METHODOLOGY AND ITS LIMITATIONS

When the need for effectiveness evaluation is recognized and emphasized, the next question is how to do it. One popular and straightforward strategy for steering effectiveness evaluation is known as the replication of efficacy evaluation strategy. Replication strategy uses the same framework and methodology used in an efficacy evaluation to evaluate an intervention program in the real world. Because of the real-world settings, it usually cannot follow all of the same procedures as an efficacy evaluation. The replication strategy advocates the idea of applying as many of the efficacy evaluation's research principles and procedures as possible when conducting an effectiveness evaluation. Under this strategy, the standard used to judge the quality of an efficacy evaluation is also used to judge the quality of an effectiveness evaluation.

However, the dynamic of an intervention in the real world is quite different from one in an idea setting. Being forced to use the framework and methodology of one evaluation on the other is likely to create problems or even chaos. Any of the following five common problems can then arise.

1. Ignoring Stakeholders' View and Buy-In

During an ideal efficacy evaluation, evaluators take the role of objective assessors, and stakeholders such as counselors and clients are motivated and enthusiastic participants. Stakeholders' buy-in already exists, which is often not the case in effectiveness evaluation. But real-world effectiveness evaluation relies on stakeholders' active collaboration. Thus, evaluators cannot afford to *overemphasize* their own objectivity, to the point of avoiding interaction with stakeholders (understanding their views, exploring evaluation issues with them). Stakeholders encountering such reserve may wonder whether the evaluator has some hidden agenda. Suspicion can discourage cooperation and even lead to rejection of the evaluation. An example is found in one assessment undertaken to determine who used a community park. Evaluators began showing up in that park daily and recording the types of people there and the activities they pursue. However, no one from the evaluation team communicated with community leaders or nearby residents about the purpose of the data-collecting effort. The stakeholders consequently felt uncomfortable with the evaluators' daily presence. It was rumored that the evaluation was, in fact, a move by authorities to manufacture some reason to close the park. Soon, a public outcry demanded the evaluation be stopped.

2. Taking for Granted the Cooperation of Implementers

When performing efficacy evaluation, evaluators get used to enjoying implementers' cooperation. When they later participate in effectiveness evaluation, they often assume that implementers will jump to do what the evaluators ask of them. But implementers' support is not automatic; they must be given due consideration when the evaluation is planned. Will the information gained rely, for instance, on implementers' documentation of clients' sexual histories? If so, how implementers feel about the nature of these data should be explored beforehand. It is possible that they will be troubled by the prospect of asking such personal questions. If simply ordered or assigned to perform this kind of data collection, they may not proceed cooperatively.

3. Neglecting Capacity Building

Similarly, evaluators cannot afford to assume that an implementing organization has the capacity to manage effectiveness evaluation. Whereas organizations typically affiliated with *efficacy* evaluation—universities and research firms—are indeed highly capable and efficient when it comes to data, the organizations pursuing effectiveness evaluation are more diverse and regularly demonstrate lesser capacities. It is so easy to take for granted that all organizations can readily provide accurate information for use in evaluation. But many organizations require coaching to prepare for this role. As an example, evaluating a media campaign's antismoking effect required hospitals in a community to share their records of identified smokers with evaluators. An organization groomed for, or experienced with, research should not find such sharing difficult. But not all hospitals fit this description. The evaluators in the example received the requested information, but soon found that a large proportion of the individuals named in hospital records actually did not use tobacco. The mistake occurred because at least one hospital had data systems unable to produce accurate information on the topic. Setbacks such as unreliable data can be avoided by considering early on whether an implementing organization merits capacity building.

4. Failing to Anticipate Environmental Intrusions

Accustomed to efficacy evaluation with all its scientific controls, outcome evaluators are not always sensitive enough to environmental intrusions affecting effectiveness evaluation. Thus, they fail to take precautions that might offset such intrusions. An outreach program to serve sex workers offers an example. The researchers in charge of the project overlooked an ongoing police campaign to clear several neighborhoods of sex workers (perhaps because researchers were unaware of the campaign). These researchers directed their staff members to do outreach among sex workers in these neighborhoods and the rest of the community. A number of the unsuspecting staff members were actually arrested, and a great deal of effort was needed to convince police of their innocence.

5. Bowing to the Academic Significance
of Evaluation at Stakeholders' Expense

Moving unscathed from efficacy evaluation to effectiveness evaluation requires that a whole mind-set change. Above all, the intellectual curiosity that is paramount in efficacy evaluation must be exchanged for a steady attentiveness to stakeholders' needs. Without such care, an effectiveness evaluation's potential academic significance can divert attention from stakeholders' concerns. We easily see how evaluators could deeply value the experience of producing theoretical results significant to the academic community. In contrast, practical and immediate issues affecting programs are what matter to

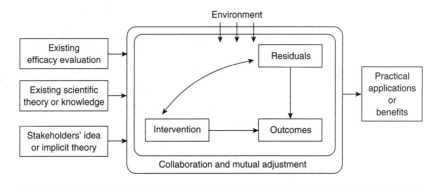

Figure 9.2 Conceptual Framework of Effectiveness Evaluation

stakeholders. As an example, consider an antismoking program at a large corporation. Evaluators were told that the stakeholders needed to know the effects of two interventions: informational letters sent to employees who smoke and antismoking seminars for employees generally. The evaluators, aware that the literature long ago established that informational letters do not work, might be tempted to design an efficacy evaluation involving only the seminar and its effect. But the stakeholders had appreciated several features of their letter: the fact that it reached all targeted individuals quickly, and was inexpensive and simple to administer. As an intervention strategy, according to the existing literature, the letter might have promised little in the way of demonstrated effect. As a practical matter, though, it promised much in the cost-effectiveness department, and this mattered to the stakeholders. Stakeholders' interests may not appeal to academic circles, but they must nevertheless give shape to the design of effectiveness evaluations.

THE CONCEPTUAL FRAMEWORK
OF EFFECTIVENESS EVALUATION

To avoid overextending the replication strategy in any of the five ways, effectiveness evaluation may need its own framework and methodology based upon the bilateral strategy to guide its application. The conceptual framework of effectiveness evaluation is illustrated in Figure 9.2.

Under the bilateral paradigm, efficacy evaluation is one of the sources for conducting effectiveness evaluation. But not every efficacy trial should have an effectiveness trial conducted on it. Only those efficacy trials with high practical implications and stakeholders' support should be pursued. As indicated in Figure 9.2, another important source is stakeholders' idea and program theory, with some preliminary data to support its plausibility. Stakeholders have accumulated broad experience in their field, so their ideas and input could be another source to set up a successful effectiveness evaluation. Figure 9.2 indicates that effectiveness evaluation requires evaluators to collaborate with stakeholders to develop a milieu that not only does not seriously disrupt normal service routine but also allows for conducting an outcome evaluation. The key to create such a milieu is mutual adjustment between stakeholders and evaluators. That is, stakeholders and evaluators recognize and accommodate each others' needs in outcome evaluation. The conceptual framework in Figure 9.2 also illustrates effectiveness evaluation's clear mandate to provide direct, practical information for stakeholders' use, as well as the need for programs to seek support from corresponding implementing organizations or, at the very least, a commitment to see the intervention implemented.

The top of Figure 9.2 highlights that effectiveness evaluation does not seal off or control the environment, which is different from efficacy evaluation. The environmental factors like to influence the intervention process. The right-hand side of the figure highlights that effectiveness evaluation has a clear mission to produce useful information for practical application. This means that effectiveness evaluation needs to admit ordinary clients and use ordinary implementers. This means that it is desirable to have interventions that are robust enough to stand the environmental intrusion but not too complex for ordinary implementers to use.

Figure 9.2 also indicates the possible correlations between the intervention and residuals that need to be dealt with. Effectiveness evaluation does not have the benefits of control and manipulation, as efficacy evaluation has, and is required to have internal and external validity at the same time. These two situations create a special challenge for evaluators to generate credible and usable information. The issues of research designs for effectiveness evaluation will be discussed in the latter sections of the chapter.

Research Example of Effectiveness Evaluation

As we have seen, Paul's (1966) efficacy evaluation provided convincing evidence that behavior therapy is superior to psychotherapy. Although his evidence further developed scientific knowledge, its import for actual practice was not immediately clear. Sloane, Staples, Cristol, Yorkston, and Whipple (1975) used *effectiveness* evaluation to test the potential real-world value of Paul's findings. They, too, compared the assessed effects of psychotherapy with those of behavior therapy, recruiting to the study some typical patients from a psychiatric outpatient clinic. Each was randomly assigned to a psychotherapy group, a behavior therapy group, or a waiting list. As in normal clinical settings, each therapist participating in the study was directed to practice either behavior therapy *or* psychotherapy, choosing the one best reflecting his or her experience. Furthermore, because therapists treated various patients for various problems, the treatment was never rigidly standardized as a single set of identical procedures. The results of the Sloane team's effectiveness evaluation showed that patients in either of the treatment groups improved significantly more than the control, or waiting list, patients. They also showed no significant difference, generally, between the outcomes of psychotherapy and the outcomes of behavior therapy. In other words, the neat, clean findings from Paul's efficacy evaluation were not generalizable to actual clinical treatment. This offered an important lesson: The findings of efficacy evaluation will not necessarily hold true in the real world.

DESIGNING AND CONDUCTING
EFFECTIVENESS EVALUATION

The conceptual framework of effectiveness evaluation (Figure 9.2) indicates the need for issues of the action model and environmental intrusion to be considered as the effectiveness evaluation is being designed. There are three large challenges. The first is the undeniable need for stakeholders' collaborative support during the process. In effectiveness evaluation, work must focus on stakeholder credibility as much as it does on scientific credibility (a dual allegiance previously discussed in Chapter 1). The second challenge is the methodology of effectiveness evaluation. Unlike efficacy evaluation, which focuses on internal validity, the methodology of effectiveness evaluation deals with both internal and external validity. Effectiveness evaluation often requires evaluators to creatively assemble a defensible, innovative design rather than apply a prefabricated design, as in efficacy evaluation.

The third challenge arising from the conceptual framework is providing stakeholders with useful effectiveness evaluation data. One mission of effectiveness evaluation is to integrate the information presented to stakeholders with pragmatic suggestions for improving programs. To understand what "pragmatic" might actually mean to particular stakeholders—in other words, to achieve a *responsive* evaluation—the evaluator must learn what stakeholders think and what they want or need. Only with this knowledge can the evaluation be designed for utility. In addition, the evaluator must be familiar with strategies that can be used to meld stakeholders' input into the effectiveness evaluation. There are some guidelines within the discipline of effectiveness evaluation that provide a means to address these three challenges.

Establish Working Relationships With Stakeholders

In effectiveness evaluation, a working relationship between evaluators and stakeholders is essential. To produce a responsive evaluation, evaluators must understand the stakeholders' perspectives, incorporating the purpose of evaluation, as well as the eventual uses for the data, in their evaluation designs. Furthermore, the execution of a high-quality outcome evaluation requires administrative support and even cooperation. This, too, calls for good communication and a working relationship between evaluators and stakeholders. The less a stakeholder is permitted to know about evaluation purposes and strategies, the more skeptical he or she will be of the evaluation and the more likely he or she is to want to impede the evaluation. *It is a myth that involving stakeholders' participation must always lessen the integrity of the evaluation.* It certainly cannot be assumed that stakeholders care nothing about the credibility of the design and methodology used in evaluating their programs. Recently, for example, a national evaluation system was developed to follow CDC-funded, health department-based anti-HIV programs. During development, stakeholders were heavily involved in shaping the scope and focus of the evaluation system. They helped determine how evaluation questions would be phrased to reflect the reality of the populations they served and also how evaluation results ultimately should be used. These stakeholders also agreed with evaluators on the value of rigorous quantitative designs (including experimental and quasi-experimental varieties) in testing their programs' outcomes (Chen, 2001).

To design an outcome evaluation, evaluators and stakeholders need, in general, a clear, mutual understanding that stakeholder input will be required and, furthermore, that stakeholder input will not extend to interference with scientific activities such as evaluation design, data collection, or data analysis.

In short, stakeholders' wishes must not compromise the work's scientific rigor. Finally, both evaluators and stakeholders should be aware of the reasons that stakeholder input is so beneficial:

- It helps evaluators better conceptualize the program.
- It makes possible the precise definition of intervention and goals (during intervention planning).
- It determines a context for use of evaluation findings.
- It prevents the substitution of evaluator preferences and values for those of local stakeholders.
- It ensures collection of information relevant to a program and useful to stakeholders.

Have Stakeholders' Input in Selecting an Intervention

The intervention that is good for efficacy evaluation is not necessary good for effectiveness evaluation. Glasgow et al. (2003) pointed out that character-istics of successful intervention in efficacy evaluation—such as intensive, complex, and highly standardized—are fundamentally different from, and even at odds with, characteristics of programs that succeed in real-world effec-tiveness settings, such as having broad appeal and being adaptable for ordinary implementers and clients. The selection of an intervention for an effectiveness trial should go beyond the interventions proposed in efficacy evaluation. For example, interventions in which stakeholders have high interest should also be regarded as good candidates for effectiveness evaluation. Even when researchers are interested in replicating an efficacy trial for an effectiveness evaluation, program designers should bring in stakeholders for consultation on the feasibility and utility of adopting efficacy intervention in place of an effectiveness evaluation.

Include Stakeholders in Deciding
Which Outcomes to Measure and How

To conduct an effectiveness outcome evaluation, one needs outcomes to evaluate; but it is not up to evaluators to arbitrarily choose the outcomes and their measures. When an evaluator has been too independent in terms of these choices, stakeholders can perceive the evaluation as irrelevant to their program, easily dismissing it. Stakeholders' input, then, also is needed when finalizing the outcomes and measures of an evaluation; that is, during *outcome specification.* The precept can be illustrated using an HIV prevention program as an example. The goal of such programs is usually to increase safe sex practices within a high-risk population; for example, Hispanic migrant workers. This goal is attractive and agreeable. The outcome measures that will be able to evaluate its fulfillment may, however, be more controversial. Scientists and funding agencies tend to define safe sex as the use of a condom in *each* act of sexual intercourse. So, when an evaluator who has not carefully thought through his or her outcome specification decides to measure safe sex with the question, "The last time you had sex, did you or your partner use a condom?" the intervention is perhaps being set up to fail—at least in the eyes of the funding agency. Why? Because if the answer to the question is no, then by agency criteria, the intervention failed, even if the client has used a condom regularly for months and did not the last time simply due to some unusual circumstance.

For many program directors and implementers within community-based organizations, a different interpretation of safe sex reigns, informed by their understanding that, due to cultural and gender factors, it is not realistic to expect high-risk individuals to use condoms for each act of intercourse. Among Hispanic migrant workers, for example, it can be deemed inappropriate, or even an insult, for a woman to ask a steady partner, such as her husband, to use a condom. HIV prevention workers in the field might well find that the scientist's or funding agency's definition renders achievement of the goal impossible. The prevention workers' own definition of safe sex might spring from a harm reduction perspective. The outcome they are likely to hope for from HIV prevention programs (at least in the current state of affairs) may be use of condoms with sex partners other than one's steady partner. Harm reduction has, in fact, been used by many community-based organizations to devise HIV prevention programs. So, an evaluation that arbitrarily took harm *elimination* as the measure of safe sex might call down on itself charges by program stakeholders that it overlooks the work the program is actually doing. At times, stakeholder groups have differing views on the outcomes of a program. By facilitating outcome specification, evaluators can help in finding a solution acceptable to all. They can prompt the HIV prevention program stakeholders, for example, to consider asking more than one question to evaluate condom use: perhaps one about the steady partner, reflecting a harm reduction orientation, and one for other partners, reflecting either the harm reduction or the harm elimination orientation.

Build Organizational Evaluation Capacity

Effectiveness evaluation operates in the real world. Evaluators rely on actual program administrators and implementers to structure a program and carry out its daily activities; these professionals also often collect the data for the outcome evaluation. The important role of program administrators and implementers means that evaluators would be wise to ensure that an implementing organization as a whole has adequate understanding and skills to support evaluation activities, and also the capacity to deliver services as intended. If either is lacking, there are ways in which evaluators can help organizations build capacity. To begin, evaluators must ascertain that the following four specific evaluation capacities exist.

1. *Capacity to Enforce Eligibility Criteria and Effectively Recruit/Screen Participants.* Outcome evaluation can become flawed if people outside the target group enroll for services. To illustrate, imagine a flu immunization program whose implementing organization intended to screen applicants and admit only high-risk individuals, like those with asthma or heart disease. However, the screening procedure authorized by this implementing organization was unreliable, and the program accepted many participants from outside its target group. Unless this shortcoming is discovered before intervention is delivered and data collected, the integrity of the evaluation will suffer.

2. *Capacity to Record Client and Service Data Precisely.* Effectiveness evaluation requires implementers and service providers to precisely record their daily activities. In terms of service delivery, four questions are vital: Who is being served? What are their social and demographic characteristics? What services are provided? What is the level of client participation? It may be a mistake to assume that implementers will be able to answer these questions. Many evaluations require, for example, data concerning the race and ethnicity of each client. It is possible, however, that data that the evaluator receives from the implementers could include a number pertaining to race that is far off the reported number pertaining to ethnicity. If this happens, the evaluator needs to facilitate an effort by implementers to build their data-collecting capacity. Similarly, evaluators often rely on implementing organizations to have the capacity to store and manage client and service data. A wise evaluator will be certain this capacity can prove itself when needed.

3. *Capacity to Collect or Assist in Collecting Outcome Data.* Stakeholders who involve evaluators in program development from the beginning will

likely enjoy real help in identifying the measurable goals they hope to achieve. Evaluators or implementers may gather the evaluation data pertaining to the associated outcome measures; if implementers take on the responsibility, then the issues of organizational capacity discussed above are again applicable.

4. *Capacity to Use Evaluation Results for Program Improvement.* One major purpose of effectiveness evaluation is to create a base of information for use in improving current or future programs. The evaluator must ask, then, if the stakeholders seem capable of deploying evaluative information in a way that will benefit their program. If they do not, the evaluator should work to build their capacity to put evaluation data to good use.

Enhance the Practical Relevance
of Innovative or Demonstration Programs

Effectiveness evaluation must have practical relevancy to the real world. Practical relevancy means that the evaluation results are relevant to stakeholders' practice in the future. In other words, stakeholders must be able to use the information to change or improve what they will be doing in the future. The concept is related to generalizability, but it provides a clear delineation of the target to which the evaluation results can be generalized. It is both stakeholders' and evaluators' responsibility to design an effectiveness evaluation with high practical relevancy.

Chen (1990) developed a generalizability strategy for enhancing practical relevancy in devising effectiveness evaluations. His strategy is represented in Figure 9.3.

The figure comprehends two systems: a research system on the left-hand side and a generalizing system on the right-hand side. The generalizing system reflects effectiveness evaluation's locus in the real world. It is within the research system that an evaluation must work. In the research system, the sample is defined, the action model designed, and the intervention implemented. In contrast, within the generalizing system lie the pragmatic, ongoing concerns of real practice, to which evaluation results will eventually be applied.

Chen's strategy increases generalizability by ensuring that the research system is designed to mimic the generalizing system as exact as possible. To craft such a design, the evaluator first needs to decide what the generalizing system's action model looks like, and then design the corresponding research system. Consider a treatment program for juvenile delinquents. The design of

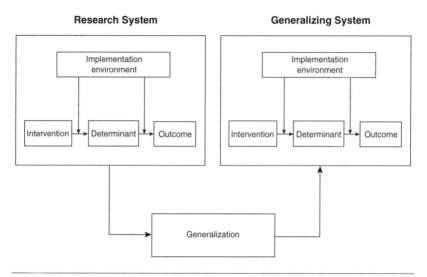

Figure 9.3   Conceptual Framework for Enhancing Practical Relevancy

SOURCE: Adapted from Chen (1990).

the program can follow two paths: one focuses only on the research system itself, whereas the other one takes into consideration both the research system and the future generalizing system. Consider an example of a program focusing only on the research system. Program designers may design a highly innovative and interesting intervention program. One youth program, for example, planned to introduce youngsters to the wilderness as an intervention discouraging juvenile delinquency. Program designers and supporters hope youngsters will learn survival skills, discovering self-discipline in the process, which they hope will reduce their criminal activities in the community. All the efforts then concentrate on finding the sites, working with the criminal justice system to recruit the juvenile delinquents into the program, training the supervisors and coaches, and getting parental and community support. In terms of the research system, the program could be sound and highly feasible. The problem is that the program may have low practical relevancy because of its failure to take the generalizing system into consideration. Self-discipline is a boon in any environment, but the concrete tasks of surviving in the great outdoors bore little resemblance to the skills that might help a young person thrive in a community.

In another example, consider an intervention that takes into consideration both the research system and the generalizing system. This intervention uses a summer camp to increase juvenile delinquents' interest in the program. However, the intervention focuses on values, knowledge, and skills that would help them stay in school, keep a good relationship with their family, stay away from gangs and criminal activities, and link them with community volunteers for ongoing support of their positive behavior. This program anticipates that the juveniles will face the same difficulties in their families, neighborhoods, and schools in the future, so the intervention is designed to help youngsters survive in the community. The practical relevance of this program is high.

Construct the Research Design

Effectiveness evaluation must produce credible evidence, just as efficacy evaluation must. However, efficacy evaluation needs to provide credible evidence to meet only the requirements of internal validity. Effectiveness evaluation needs to provide credible evidence for meeting both internal and external validity. This premise implies that the foundation for applying research designs in effectiveness evaluation cannot be identical to that used in efficacy evaluation. The fundamental issues of effectiveness evaluation are as follows.

1. Select a Research Design That Has Merits for Optimizing Both Internal and External Validity

The principle and priority for guiding selection of a research design for effectiveness evaluation is different from that for efficacy evaluation. Randomized controlled experiments are the most rigorous design in efficacy evaluation. The issue is what the role of the randomized experiments will be in effectiveness evaluation. This issue needs to be discussed at both a technical and a utility level. At the technical level, it is extremely difficult to apply randomized controlled experiments to effectiveness evaluation because it is impossible to impose tight control over the real world. However, it is possible to apply "randomized experiment" in effectiveness evaluation. The randomized experiments focus mainly on random assignment of clients or participants into intervention and comparison groups. Theoretically speaking, even without controlling the environment or recruiting the homogeneous group of clients, the randomized experiments are still the most powerful designs in achieving internal validity in effectiveness evaluation (Shadish et al., 2002). Randomized experiments are more difficult to apply in the real world than are other designs, However, this is not a justifiable reason against the use of randomized experiments in effectiveness evaluation. Boruch (1977) provides many examples of applying randomized experiments in effectiveness evaluation.

In the utility level, an important issue is whether randomized experiments should be regarded as the most desirable or ideal design in effectiveness evaluation. Or, should randomized experiments be regarded as one of the optional designs to be considered? Because effectiveness evaluation needs to meet the requirements of both internal and external validity, randomized experiments may not always be the optimal design.

William S. Gosset's pioneering work on effectiveness evaluation illuminates this issue. Gosset, the eminent statistician who formulated the t-test (and used a pen name, "Student") (Student, 1936a, 1936b), challenged Ronald Fisher's argument that randomized experiments offered the best design for agricultural research. Gosset argued that because plots needed to be artificially arranged to meet randomization's requirements, the estimate of standard error derived from randomized experiment would fail to reflect the standard error as derived in a natural setting. Agricultural research is ultimately intended to benefit real farmers, he continued. Wouldn't the best design, then, be one that bore direct implications for farmers' work? Gosset devised such a design, an effectiveness evaluation or trial (equivalent to a research system) that mimicked farmers' actual practice (equivalent to a generalizing system). He called it the "half-drill strip" design. Under this design scheme, agricultural plots were assigned to treatment varieties A and B and arranged in the pattern ABBAAB-BAABBA. Gosset argued that, in this way, inherent errors such as linear trend could be adequately dealt with—and by average farmers themselves! To set up a half-drill strip trial, they simply filled the seed boxes on the left side of the seed hopper with treatment variety A and those on the right side with treatment variety B. Then, the horse or tractor moved down one row sowing seeds in AB order. The row completed, the horse or tractor turned. With the seed hopper positions unchanged, it made its return trip sowing the two treatment varieties in BA order this time. Such a procedure was admittedly less rigorous than randomized controlled experiment, Gosset wrote, but it had the attractive feature of resembling farmers' actual practice. Therefore, research results obtained using the procedure were easily generalized to practical use. Fisher (Barbacki & Fisher, 1936) was one of Gosset's contemporaries who was not convinced. He insisted that, from a statistical viewpoint, without randomized assignment of treatment varieties, the half-drill strip resulted in an inaccurate, invalid estimation of standard error.

The debate offers some important lessons. Outcome evaluation design ideally maximizes the degree of precision with which an intervention or treatment effect is estimated. But as the saying goes, there is no free lunch. To achieve maximum internal validity of the evaluation, heed Fisher's objections to the half-drill strip method and choose traditional efficacy evaluation, even though it ignores external validity. To achieve dual validity, on the other hand, follow Gosset. The effectiveness trial sacrifices a degree of internal validity, which makes it possible to obtain external validity. It is a trade-off. Of course, neither Gosset nor this book intends to imply categorical rejection of randomized experiment in effectiveness evaluation. The important point is that randomized experiments are not, de facto, the most appropriate. Evaluators and stakeholders need to assess the nature of the program and its circumstances to select or develop an optimal design—quantitative, qualitative, mixed—for effectiveness evaluation.

2. Apply Prefabricated or Synthetic Designs

Both experimental and quasi-experimental designs (Cook & Campbell, 1979; Shadish et al., 2002) are options to be considered in the effectiveness evaluation. Methodologically speaking, quasi-experimental designs cannot compete with randomized experiments in achieving internal validity. However, quasi-experimental designs have advantages in external validity because they better fit the natural process of the real world. The selection of which design to use in effectiveness evaluation is not as straightforward as in efficacy evaluation. The selection of a randomized experiment in effectiveness evaluation depends on factors such as whether the randomization creates too much disruption on the routine administrative work. For example, if a program is in high demand and clients need to wait for intervention, the use of randomization to determine who should get the intervention is highly justified. Another factor is the willingness of stakeholders to trade external validity for internal validity. If they demand high internal validity for reducing controversies and do not mind the loss of external validity, then randomized experiments are a good choice. Otherwise, quasi-experimental designs are more suitable for effectiveness evaluation.

Randomized experiments are prefabricated designs, as are quasi-experimental designs such as time-series designs, nonequivalent control group designs, and regression discontinuity designs. Prefabricated designs are designs with all of the elements placed in a systematic package. As long as evaluators and researchers follow closely the procedure described by the design, they will achieve high internal validity. The prefabricated designs simplify evaluators' task. However, intervention programs operate in the real world and vary a lot in terms of evaluation circumstances. A prefabricated design may be insufficient for meeting evaluation needs. Under this condition, evaluators need to devise a design for the evaluation circumstances. This kind of design often combines a few design elements or a few designs together into a new design. This kind of design is called a patch-up design (Cordray, 1986) or a synthetic design (Chen, 1990). The following strategies are frequently used in the construction of patch-up or synthetic designs.

Pattern Matching. A method called *pattern matching* can help. Pattern matching is a detailed elaboration of theoretical patterns between intervention and outcomes for the purposes of empirical testing. When patterns are confirmed to exist, the validity of any conclusions about the program's effect is increased. Pattern matching is analogous to comparing fingerprints. As argued by Trochim and Cook (1992) and Mark, Hofmann, and Reichardt (1992), the more detailed the elaboration of patterns, the better the fit of theoretical outcome data to observed outcome data—and the greater the evaluators' confidence in their assessment of the program's effectiveness.

Pattern matching is particularly useful when a control or comparison group for an outcome evaluation is unavailable. A challenge, however, must be met: the identification of an outcome variable that (a) is not targeted by the intervention, and (b) is subject to the same disturbances as the target outcome variable. If the changes are observed in both the target and nontarget outcome variables, this would, in theory, mean that the disturbances rather than the intervention had affected both. If the changes are observed only in the target outcome variable, then they are unlikely to have resulted from rival hypotheses or disturbances. Trochim (1984) illustrated pattern matching with the following example. If an algebra tutorial program for disadvantaged students has an effect, then in theory, the students' algebra scores should rise, but their scores in other subject areas, such as geometry, should not rise—at least not in the short term. Matching the score/subject pattern for algebra to score/subject patterns for other subjects should help solidify causal inferences for the effect of the algebra tutoring.

The rationale for pattern matching is the theoretical principle that all related outcome variables should experience identical influences from any confounding factors or disturbances. Thus, a student's scores for various mathematical subjects all should be identically influenced by his or her maturation, motivation for achievement, and experience of events (e.g., decreased parental involvement). Assuming the similarity of the influences, a pattern marked by a sudden increase only in the score targeted by intervention, with no increase in other scores, indicates the likely confirmation of the intervention's effect.

Pattern matching can be combined with a prefabricated quasi-experiment to create persuasive evidence of program effectiveness. For example, McKillip (1992) combined pattern matching with an interrupted time series design to evaluate a health promotion campaign. The purpose of the campaign was to encourage responsible attitudes and behaviors concerning alcohol in a university setting. Time series data on three outcomes—responsible alcohol use, good nutrition, and stress reduction—were collected before and after the intervention. The targeted outcome variable was responsible alcohol use, so the nontargeted outcome variables were good nutrition and stress reduction. Theoretically speaking, if the campaign was exhibiting an effect, measures for responsible alcohol use should change, whereas those for good nutrition and stress reduction should not. The study found that observed evaluation patterns in fact matched the theoretical patterns very well.

Mixed Methods. Evaluation validity may also be enhanced by the strategy of applying mixed methods, a combination of qualitative and quantitative procedures (e.g., Greene & Caracelli, 1997; Yin, 2003). Quantitative and qualitative research methods each have strengths and weaknesses, so combining them in a broader inquiry produces more and deeper insights into a program's effectiveness. For example, Weiner, Pritchard, and Frauenhoffer (1993), successfully combined qualitative and quantitative methods to evaluate an anti-drug use program, obtaining qualitative and quantitative data evidencing the program's effectiveness.

ADDRESSING DIFFERENTIAL
IMPLEMENTATION OF INTERVENTIONS

An intervention whose outcomes are to be evaluated needs a structure that allows for comparisons. By structuring an intervention so that one group receives the intervention and another does not (perhaps receiving a different intervention), the program designer is allowing for comparisons to be made between the two groups' outcomes. Intervention structures that vary the implementation level or intervention components for different groups, regions, or times are called *intervention conditions.* Precise intervention conditions can be factored into the research design. When classic controlled experiment is selected by the evaluator, the intervention conditions will be scrupulously manipulated via use of an experimental group and a control group. Using such an efficacy technique, researchers improve the odds that the integrity of the intervention can be maintained throughout the experiment. Nevertheless, even in efficacy evaluation, the implemented program may prove to deviate from the intervention conditions (Shadish et al., 2002).

This problem is worse for effectiveness evaluation, in which evaluators have little direct control of the implementation environment in which an evaluated intervention is delivered. Implementation levels commonly change with the individual, the group, or the area. Also, so-called control group or comparison group members are absolutely free to receive some other intervention. Further complicating the work, it is possible that clients in the intervention group and those in the control or comparison group will interact with each other regularly. For these reasons and others, actual intervention conditions tend not to mirror planned intervention conditions exactly. Evaluators should be able to employ capacity building to attempt to bring consistency to the two. It will remain necessary, however, to measure the level of their fidelity, for in effectiveness evaluation, no assumption can be held about the consistency and integrity of an intervention's implementation. (Responding to this reality, additional process evaluation techniques are deployed in effectiveness evaluation, including fidelity assessment.) Upon determining that the implementation of an intervention has been precise, evidenced by the general fidelity of the actual intervention to the program model, the evaluator can proceed to analyze the relationship between intervention and outcome.

Measuring levels of intervention implementation for incorporation in the outcome assessment is crucial. In theory, if an intervention is effective, clients who attend every session, or otherwise receive the entire intervention package, should demonstrate better outcomes than clients who do not. The work of McHugo, Drake, Teague, and Xie (1999) on the assertive community treatment model provides an example. Studying fidelity between assertive community treatment and client outcomes, the team found that, across programs, treatment was delivered with varying levels of fidelity to the model. Furthermore, the researchers established that clients in programs with greater fidelity showed greater reductions in their substance abuse (attaining a higher rate of remission for substance-use disorders) than did clients in programs with less fidelity.

ANTICIPATING INTRUSIONS
AND PREPARING FOR PROBLEMS

In the real world, where effectiveness evaluation operates, unexpected disruptions can create problems for a program and for the evaluation of its outcomes. Stakeholders can do nothing about some problems, such as natural disasters. Other problems may be alleviated through brainstorming by evaluators and stakeholders about vulnerable areas of the program, and subsequent development of countering strategies. If an education program, for example, was built around an innovative new textbook for its classrooms, and if the publishers were falling behind in producing the books as the stakeholders worked through the program planning stage, it would be very reasonable for an evaluator to raise the question of what would happen if the textbooks were not available on time. Program staff and evaluators would have to work together at devising a strategy to address this issue.

BEYOND THE BASIC
EFFECTIVENESS AND EFFICACY EVALUATIONS

The above discussion is a basic effectiveness evaluation based upon the conceptual framework of effectiveness evaluation in Figure 9.2. The focus of this framework is to assess intervention outcomes in the real-world setting. However, in many situations, for facilitating practical application or adoption, potential users or adopters may need to know what it has taken to reach these outcomes. Under this condition, a comprehensive effectiveness evaluation is needed to provide such a holistic assessment. A comprehensive conceptual framework for effectiveness evaluation in shown in Figure 9.4.

Figure 9.4 indicates that a basic effectiveness evaluation can be expanded to provide the following information to enhance its practical relevancy.

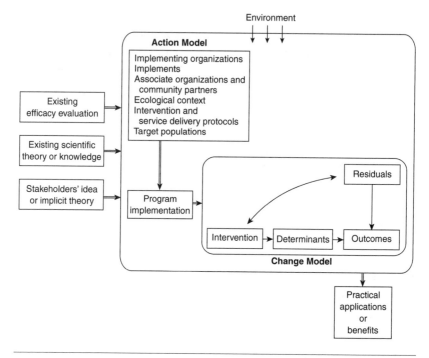

Figure 9.4   Comprehensive Conceptual Framework of Effectiveness Evaluation

How the Program Is Implemented

For practical use or adaptation, potential users and adopters need to know how and what it has taken to reach the intervention outcomes. A typical effectiveness evaluation can be expanded to include an evaluation of the implementation process. Figure 9.4 indicates that evaluators can evaluate systematically how the action model is actually implemented.

How the Underlying Causal Mechanisms are Operated

Potential users or adopters will be more confident about the intervention outcomes of the program if the evaluation can illustrate how the causal mechanisms underlying the program are operated. Figure 9.4 indicates that a basic effectiveness evaluation can be expanded to evaluate the change model to address this issue.

How to expand an effectiveness evaluation to assess implementation processes and/or underlying causal mechanisms is a topic of theory-driven outcome evaluation. This topic will be discussed in detail in Chapter 10. Similarly, an efficacy evaluation can be expanded to a theory-driven outcome evaluation in order to address the issues of how and why an intervention succeeds or fails to reach its goals. These issues will also be covered in Chapter 10.

TEN

THEORY-DRIVEN
OUTCOME EVALUATION

———•◦•———

Traditionally, discussions of outcome evaluation, either efficacy or effectiveness, have focused heavily on assessing the relationship between intervention and outcome. Chapter 9's discussion of efficacy evaluation follows this custom. Effectiveness evaluation, also a topic of Chapter 9, is equally accustomed to focusing on the intervention-outcome relationship during the actual assessment (although effectiveness evaluation can be enriched when the framework of program theory is applied to it for the purpose of structuring a real-world program; in other words, contextual information is not integrated into the evaluation process). If an outcome evaluation's primary tactic is assessing the relationship between intervention and outcome, it is a *black box* or *input-output evaluation.* Black box outcome evaluation, whether efficacy or effectiveness, does not systematically evaluate transformation processes that turn interventions into outcomes. Instead, black box evaluation largely searches out information about a program's merits. When stakeholders ask only for unequivocal evidence of an intervention's gross effect on predetermined goals, black box evaluation is sufficient. However, if stakeholders and evaluators desire to understand both the merits of a program *and* how its transformation processes can be exploited to improve the intervention, then *theory-driven outcome evaluation* is often a better choice. Theory-driven outcome evaluation does more. (Many call it simply *theory-driven evaluation,* or sometimes *theory-based evaluation.* To maintain a distinction from the theory-driven process evaluation explored in Chapter 7, however, I will use theory-driven outcome evaluation.) Theory-driven outcome evaluation takes into account both underlying causal mechanisms and the implementation process when assessing the effect of a program. It can provide stakeholders with understanding of whether a program is reaching its goals and document insightfully the hows and whys of program success or failure (Bickman, 1990; Chen, 1990; Weiss, 1997). This is why theory-driven outcome evaluation has a place in the taxonomy under enlightenment assessment strategy (see Table 3.1): It enlightens stakeholders concerning the crucial assumptions under which their program is expected to operate in the field each day and concerning the program components that contribute to, and hinder, overall success. This chapter will show how efficacy or effectiveness evaluation can, using program theory, be expanded or upgraded into a theory-driven outcome evaluation. Neither efficacy nor effectiveness evaluation has to follow the black box evaluation route.

The advantages cited here for theory-driven outcome evaluation—as well as other advantages—are well documented in the literature (e.g., Bickman, 1990; Chen, 1990; Donaldson, 2003; Weiss, 1997). Nevertheless, a brief review of the three most significant advantages should be helpful at this point.

1. *Theory-driven outcome evaluation serves accountability and program improvement needs.* In general, so-called black box evaluation fulfills accountability requirements only. The information provided by theory-driven outcome evaluation can be used to serve accountability and program improvement needs alike because theory-driven outcome evaluation generates *two* kinds of information for stakeholders. Initially, it assesses whether a program is achieving its predetermined goals, which meets stakeholders' need for accountability; in this, it is like black box evaluation. Subsequently, it investigates why and how a program succeeds or does not succeed, fueling stakeholders for the task of better understanding and improving their programs. By investigating underlying causal mechanisms, theory-driven outcome evaluation has potential to contribute to substantive knowledge or science in general.

2. *Theory-driven outcome evaluation can comment on construct validity.* Construct validity measures the degree to which an outcome evaluation assesses the right thing in the right way. As explained in Chapter 1, it is sometimes possible for programs to achieve their goals via several different channels. Some of these channels, however, may be illegitimate or may lack social approval. A black box evaluation performed on a program that succeeds via unsanctioned channels will ignore underlying causal mechanisms and so miss

this very obvious demerit. A program that achieves its effect in illegitimate or socially disapproved ways can hardly be termed a success. Thus, the black box evaluation may be ineffective at examining construct validity. Theory-driven outcome evaluation has higher construct validity because it examines causal mechanisms underlying a program, which can provide evidence that the evaluation is assessing the right thing. For example, an education program claims that its intervention program can increase students' reading scores by enhancing students' motivation to read. We are more confident of an outcome evaluation that shows that, indeed, students' motivation was improved and students' reading scores were increased than of an evaluation that shows only that the program was successful but provides no information on whether students' motivation was increased.

3. *Theory-driven outcome evaluation can increase internal validity.* The theory-driven outcome evaluation requires specification of program theory. The process of specifying program theory adds the advantage of theory-driven outcome evaluation to apply pattern matching, as discussed in Chapter 9. As pointed out by Trochim (1998), Pawson and Tilly (1997), and Chen (1990), elaboration of theoretical patterns can enhance internal validity.

A REPLY TO CRITICISMS OF THEORY-DRIVEN OUTCOME EVALUATION

In spite of the popularity of theory-driven outcome evaluation, scholars claim to have identified certain flaws in it (Scriven, 1998; Stufflebeam, 2001). It is well worth considering these scholars' basic criticisms, as well as responses addressing them. Some are briefly presented here. To begin, Scriven (1998) professes that the task of program evaluation is to assess the merit or worthiness of a program, which is an activity indistinguishable from evaluating products such as dishwashers and automobiles. Just as in the work of *Consumer Reports,* the credibility of program evaluation rests on evaluators' objectivity. Because there is no need to understand how and why a dishwasher works in order to assess its merit, there is no need, either, to understand how and why a program works in order to assess its merit. This book has, from the start, established a different viewpoint. What a program evaluation is and what a program evaluator does far exceed the limited role assigned by Scriven. The whole of the reasoning supporting this argument is available in Chapter 1, but its larger points are summarized here.

Scriven ignores human factors in program evaluation. Product evaluators do not need to concern a dishwasher's or automobile's views; program evaluators have to take stakeholders' views and concerns into consideration in order for an evaluation to be considered credible and fair. Scriven's model narrowly focuses on assessing the intrinsic merit of a program; for example, it may be justifiable to assess a television based mainly upon its intrinsic features, such as the quality of picture, sound, durability, style, and so on. However, an evaluation of a program needs to consider both intrinsic and extrinsic factors. Because human beings highly react to both program participation and evaluation, the merit of a program cannot be assessed adequately without an awareness of how the accomplishment of goals has been pursued—Scriven's model for program evaluation could be misleading. Furthermore, in contrast to product evaluation, program improvement is commonly a central purpose of program evaluation, and Scriven's limited view of program evaluation does not fit this reality. As discussed in Chapter 1, an effective practical evaluation must be future action-directed, have both scientific and stakeholder credibility, and take a holistic approach. Scriven's model and arguments lack these important features; however, the theory-driven outcome evaluation model could overcome these shortcomings. The soundness of Scriven's criticisms of theory-driven outcome evaluation is highly questionable.

Stufflebeam (2001) also does not recommend theory-driven outcome evaluation. However, his primary criticisms, which follow, inaccurately reflect the nature and procedures of theory-driven outcome evaluation. He says that

1. Evaluators using program theory might, in fact, develop their own theory and evaluate it, creating conflict of interest.

2. Theory-driven outcome evaluation depends on a base of sound theories, which few programs can point to as their foundation.

3. Evaluators using theory-driven outcome evaluation can tend to displace whatever program staff members have been using to create the program design.

4. Theory-based outcome evaluation is too difficult to do correctly.

Stufflebeam's criticisms plainly misconstrue authentic program theory. For example, concerning his first objection, program theory is actually *stakeholders'* program theory. Stakeholders have made crucial assumptions, and evaluators' role is to make those assumption explicit in order to allow for the assessment of stakeholders' programs. To my knowledge, no one advocates the imposition of evaluators' own agendas on stakeholders' programs, which they then evaluate. Nor are there examples of evaluations that succumbed to such a practice. Stufflebeam's second criticism has been addressed in Chapter 2. I explained there that a program to be evaluated can be based on explicit, validated scientific theory *or* on stakeholders' implicit theory. In either case, theory-driven outcome evaluation is applicable, and its success will be entirely unrelated to the philosophical source of the program. Theory-driven outcome evaluation builds mainly upon evaluators' ability to help stakeholders document an explicit program theory and systematically implement it.

Stufflebeam's third criticism does incorporate one point not to be disputed: Program design is the program staff's job. However, assisting or facilitating does not equate with usurping, as professional program evaluators are aware. Thus, in actuality, through theory-driven outcome evaluation, stakeholders are urged to design and implement jobs better and more thoroughly using the comprehensive and meaningful information made available to them in evaluation results. Finally, Stufflebeam's stance on the difficulty of conducting theory-driven outcome evaluation correctly has not been substantiated with any evidence. To the contrary, the discussion in this chapter will describe the application of theory-driven outcome evaluation approaches, citing examples in which, through correct application of program theory, the associated procedures are carried out without great difficulty, resulting in very useful conclusions.

ISSUES ON THE DIFFICULTY OF CLARIFYING STAKEHOLDERS' IMPLICIT THEORY

Chapter 2 explained that when scientific theory is the basis of a program, that theory usually specifies appropriate determinants. For example, if based on the health belief model (see Kohler, Grimley, & Reynolds, 1999, as an illustration), a program is "assigned" the determinants *perceived susceptibility, perceived severity, perceived benefits, perceived barriers,* and *efficacy expectations;* each must be included in any model of the program's intervening mechanisms.

Because a program theory is, in such cases, already in place, designing and conducting theory-driven outcome evaluation of the program is a straightforward prospect. Admittedly, most programs do not spring from well-developed scientific theory. The majority have their sources in stakeholder theories, a fact that critics of theory-driven outcome evaluation have seized to submit that evaluators will meet with insurmountable obstacles should they attempt to help stakeholders clarify (or develop) their theories.

However, the Laub, Somera, Gowen, and Diaz (1999) study offers a concrete, highly persuasive example that invalidates critics' dismissal of theory-driven outcome evaluation because of purported obstacles. Laub's team considered HIV prevention programs for youth conducted by community-based organizations (CBOs). The great majority of these are based upon stakeholder theory, not scientific theory. Interventions popular with CBOs for HIV prevention are classes, HIV-positive speakers, volunteerism, condom distribution, and outreach. The Laub team's work traces each of these popular options to implicit stakeholder theory that was made explicit. The following paragraphs discuss a few examples of their study.

The first example is the intervention activity to provide HIV prevention information in the form of an AIDS 101 class, a quiz show game, pamphlets, and fact sheets. These interventions do not come from nowhere, but rather from stakeholders' theory. Their implicit theory is that youth do not have accurate information about HIV, which is why they have unsafe sex. Accordingly, if these youth are provided with accurate information, they are more likely to practice safe sex. Stakeholders' implicit theory can be made explicit, as illustrated in Figure 10.1a.

The second example is a presentation by an HIV-positive speaker. Stakeholders' implicit theory for having such an intervention is that youth have never known a person their age with HIV, so they feel invincible. Stakeholders' implicit theory can be made explicit, as illustrated in Figure 10.1b.

According to Figure 10.1b, stakeholders' theory is that presentations by HIV-positive speakers will demonstrate to youth that they are not invincible to HIV disease, which will increase their practice of safe sex.

The third example is service and volunteerism, such as participating in an AIDS walk, visiting AIDS hospices, and delivering meals. Stakeholders' implicit theory for this intervention is that youth lack compassion for people living with HIV/AIDS and do not feel vulnerable to HIV. Their implicit theory can be made explicit as illustrated in Figure 10.1c.

As shown in Figure 10.1c, stakeholders believe that service and volunteerism can increase youths' compassion for people living with HIV/AIDS and show them that they are vulnerable to HIV, which, in turn, will increase their practice of safe sex.

Laub et al. (1999) argue that making a program theory explicit has another advantage for stakeholders: They become aware of the potential limitations of an intervention. For example, in the example of providing HIV information, the theorizing process might indicate that accurate information alone might not

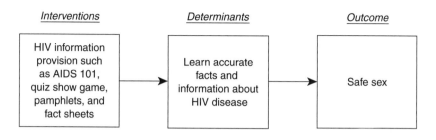

Figure 10.1a   Implicit Theory for HIV Prevention Information Intervention

SOURCE: Adapted from Laub et al. (1999).

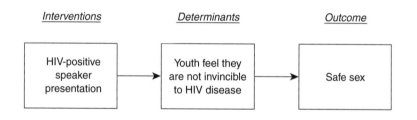

Figure 10.1b   Implicit Theory for Presentations by HIV-Positive Speakers

SOURCE: Adapted from Laub et al. (1999).

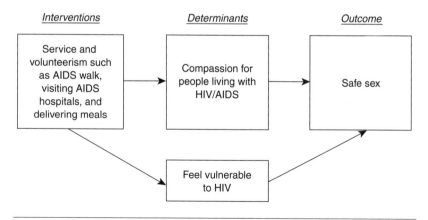

Figure 10.1c   Implicit Theory for Service and Volunteerism Intervention

SOURCE: Adapted from Laub et al. (1999).

be sufficient to motivate safe-sex behavior. This kind of information is useful for stakeholders to consider when adding other components to strengthen their program.

As suggested in Chapters 3 and 4, evaluators can use interviews or working group meetings to help stakeholders pin down their theories and especially to identify determinants in their change models. For a small program, with good preparation it can take only one or two meetings to accomplish this. For a large program, more meetings will probably be needed. When implicit theories have been clarified or made explicit, specifying the change model used for theory-driven outcome evaluation is easy.

ISSUES ON PROGRAM THEORY CONSENSUS AMONG STAKEHOLDERS

Agreement among stakeholders about program theory is often not difficult to reach. However, even if some components of the program theory do spark disagreement among key stakeholders, it is not an obstacle to evaluation. Rather, disagreement means there may be a need for testing stakeholders' competing hypotheses in the evaluation. For example, if stakeholders disagree on which determinant mediates between the program's intervention and its outcome, the evaluator could gather empirical data on the relative importance of competing determinants and provide that feedback to stakeholders. Using the example depicted in Figure 10.1c, key stakeholders may disagree on what mediates the program's intervention (volunteer activities) and the program's outcome (safe sex). Two competing determinants are considered: increased compassion for people living with HIV/AIDS and the participants' elevated feelings of susceptibility to HIV. Evaluators could include both hypotheses in the evaluation and provide empirical data that would demonstrate the relative importance of both. The same argument can be applied to situations where stakeholders disagree on other components, such as intervention activities or outcomes.

GUIDELINES FOR CONDUCTING THEORY-DRIVEN OUTCOME EVALUATION

There are three general guidelines available to evaluators who are designing and conducting a theory-driven outcome evaluation.

1. *Establishing Common Understanding Between Stakeholders and Evaluators of What Theory-Driven Outcome Evaluation Is and What It Does.* Both parties must agree on the need for such evaluation, the steps to be used in the evaluation, and the defined roles each party will play in the evaluation. The stakeholder-evaluator dialogue should specifically answer the following questions to everyone's satisfaction: What will the evaluative information be used for? Do stakeholders want an efficacy evaluation or an effectiveness evaluation? What is the program theory? What procedures could be used to clarify the program theory or help stakeholders develop one?

2. *Clarifying Stakeholders' Theory.* Program theory is the foundation of a theory-driven outcome evaluation. Evaluators could use the conceptualization facilitation approach discussed in Chapter 4 to clarify stakeholders' program theory or help them develop one. An early question for stakeholders and evaluators is whether evaluation should focus on a change model, an action model, or both. The answer reflects the extent of evaluative information desired by the stakeholders. Typically, theory-driven outcome evaluation focuses on the change model (program rationale) by examining the relationships among intervention, determinants, and outcomes. At times, however, program stakeholders wonder how program components (i.e., type of implementers, delivery models, clients) relate to the intervention-outcome relationship. Such an inquiry relies on findings about an action model (program plan), a topic discussed in Chapter 5. Alternatively, both change model *and* action model come to the forefront when stakeholders express a need for outcome evaluation combined with process evaluation. These different types of theory-driven outcome evaluation are detailed in the remainder of this book.

3. *Constructing Research Design.* The third and final guideline directs research design and data collection. Theory-driven outcome evaluation demands ample contextual information and therefore tends to rely on mixed methods of data collection. The framework of program theory meaningfully links quantitative and qualitative data in order to generate a holistic view of a program (Chen, 1996). In its comprehensiveness, a theory-driven outcome evaluation comes to rely on the collection of data beyond that typifying black box evaluation. For example, in preparation for assessing a change model, a theory-driven outcome evaluation needs to gather data on the intervention and outcome (as black box evaluation does), and also on the determinant. Collecting additional data can mean increasing costs, but these are minimized when theory-driven outcome evaluators—fulfilling their tasks of site visits,

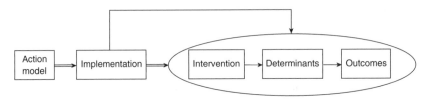

Figure 10.2   Elements of Theory-Driven Outcome Evaluation

client surveys, or other field-based collection of intervention and outcome data—arrange to collect the additional information concurrently.

TYPES OF THEORY-DRIVEN OUTCOME EVALUATION

Based upon the conceptual framework of program theory as shown in Figures 2.1 and 2.2, the basic elements of theory-driven outcome evaluation are illustrated in Figure 10.2.

The illustration makes clear that the core of theory-driven outcome evaluation is assessment of relationships among the intervention, the determinants, and the outcomes. These relationships, however, are influenced by the implementation of the action model; that is, by whether that implementation is of poor or high quality. For example, when implementers are incompletely committed to implementing the intervention, it is not likely that their intervention will much affect the determinants or outcomes. Figure 10.2 represents the two links between program implementation and the change model. One exists in the implementation that activates the intervention (the double-bent arrow). The implementation delivers the intervention to clients. The other link exists in the implementation process that shapes causal relationships among variables in the change model (the solid arrows from implementation to change model).

To be more specific, two kinds of causal mechanisms may underlie a program: mediating and moderating. A mediating causal mechanism is a component of a program that intervenes in the relationship between two other components. Figure 10.2 illustrates a determinant mediating the relationship between an intervention and an outcome. In a relationship of this type, the intervention cannot affect the outcome unless it also affects the determinant. But simply affecting the determinant does not guarantee the intervention's success, which also requires the determinant's ability to change the outcome. Thus, selection of the appropriate determinant is central to a program's performance.

For example, consider a program offering HIV prevention classes to female migrant workers. This intervention suggests that program designers believe that female migrant workers court risk of infection because they lack knowledge of HIV/AIDS and the skills to avoid the virus. Education—the class—is the program's intervention. Its determinants are HIV/AIDS knowledge and prevention skills. Its measurable outcome is condom use. Thus, to be effective, the educational intervention must alter the women's knowledge and skills; furthermore, the program designers' belief that lack of knowledge and skills causes the women's high-risk behavior must be valid. Again, the determinant mediates, or is an intervening variable in, this causal process.

The second type of causal mechanism—moderating—represents a relationship between program components that is enabled, or conditioned, by a third factor. In the presence of this third factor, the two components' relationship is manifested. In its absence, this relationship dissolves. Returning to the HIV prevention program to illustrate, assume that stakeholders suspect that their program's effect on the female migrant workers hinges on their sexual partners' cultural backgrounds. The stakeholders worry that if those backgrounds prize male dominance, the intervention will accomplish nothing. On the other hand, without cultural validation of male dominance, the intervention will likely succeed.

Isolating disparate parts of the framework of causal mechanism (see Figure 10.2), at least three kinds of theory-driven outcome evaluation can be designed: (a) an intervening mechanism evaluation approach, which focuses on the mediating process; (b) a moderating mechanism evaluation approach, which focuses on the moderation process; and (c) an integrative process/outcome evaluation approach, which focuses on the linkage of intervention to outcomes via implementation and causal processes.

THE INTERVENING
MECHANISM EVALUATION APPROACH

The intervening mechanism evaluation approach assesses whether the causal assumptions underlying a program are functioning as they had been projected to by stakeholders (Chen, 1990). To date, intervening mechanism evaluation is the most popular application of theory-driven outcome evaluation (Donaldson, 2003; Mark, 2003). It is not always labeled in the same way by those who apply it. Some evaluators have referred to it as "theory of change

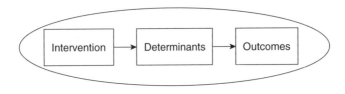

Figure 10.3   The Basic Model of Intervening Mechanism Evaluation

evaluation" (Connell et al., 1995) or "theory-based evaluation" (Rogers, Hasci, Petrosino, & Huebner, 2000; Weiss, 1997).

The basic task of the intervening mechanism evaluation is to assess the change model portion of the program theory. The intervening mechanism is modeled in Figure 10.3.

Figure 10.3 indicates that the major difference between intervening mechanism evaluation and traditional efficacy and effectiveness evaluation is the former's inclusion of determinants intervening between intervention and outcome in the model. Before going on to the other modes of intervening mechanism evaluation, a research example of intervening mechanism evaluation is presented in the hope that acquaintance with the sample case up front will simplify the remaining discussions for readers. The example is an evaluation of the Adolescent Alcohol Prevention Trial (Donaldson, Graham, & Hansen, 1994). This project focused on the three determinants said in the literature to be important causes of adolescents' drug use: skills for resisting drug use, perception of the prevalence of drug use, and cognitive acceptance of drug use. Resistance skills (i.e., the behavior involved in turning down drug offers) were taught to adolescents enrolled in the program. In addition, a normative education program helped correct the participants' erroneous perceptions about the prevalence and acceptance of adolescent substance use and sought to establish conservative norms. The evaluation design incorporated testing of the two interventions' ability to affect the determinants. It also assessed whether the determinants had the power to affect three outcome variables: alcohol use, cigarette use, and marijuana use. Figure 10.4 shows the change model for the Adolescent Alcohol Prevention Trial.

Data from the intervening mechanism evaluation showed that normative education was employed successfully in the program; it activated beliefs concerning the prevalence and acceptance of drug use, which in turn reduced drug use. The data also showed that, although resistance training did strengthen the adolescents' resistance skills, possessing such skills did not affect drug use.

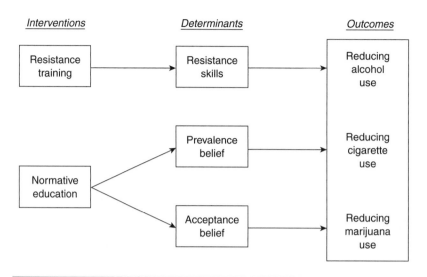

Figure 10.4   The Change Model for the Adolescent Alcohol Prevention Trial

SOURCE: Adapted from Donaldson et al. (1994).

The authors argued that their results strongly support the theoretical underpinnings of normative education interventions. Their study demonstrates the kind of information that theory-driven outcome evaluation provides: It answers whether an intervention affects outcomes, and then answers *why* the intervention does or does not do so.

Two Models of the Intervening Mechanism Evaluation Approach

Two basic models of intervening mechanism evaluation predominate in the discipline: linear and dynamic.

The Linear Model

The linear model is currently a very popular application of intervening mechanism evaluation. Linear models assume that the causal relationships among interventions, determinants, and outcomes are unidirectional: intervention affects determinant, and determinant then affects outcome. No reciprocal relationships operate among the variables. In linear models, the number and sequence of the determinants under study determine the model's form. The following causal diagrams illustrate the common linear model forms.

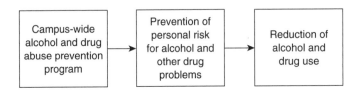

Figure 10.5   An Example of a One-Determinant Model

SOURCE: Adapted from Miller et al. (2000).

One-Determinant Model. This model, represented by Figure 10.5, contains a single determinant and is the fundamental model for intervening mechanism evaluation.

The one-determinant model is illustrated here by an evaluation of an alcohol and drug abuse prevention program at a college (Miller, Toscova, Miller, & Sanchez, 2000). The intervention consisted of multiple components: print media, videotapes, speakers, referral services, and development of self-control. The determinant was perception of risk, and the outcome was a reduction in alcohol and drug use among the students on the campus where the program was established. As predicted, the data showed that after the interventions, there was heightened awareness on campus of the risks of substance abuse, which in turn reduced alcohol and drug use there. The one-determinant model is relatively easy to construct. For example, all of the stakeholders' implicit theories in HIV prevention activities illustrated in Figures 10.1a, 10.1b, and 10.1c are one-determinant models.

Multiple-Determinant Model, No Sequential Order. Another common linear model is the model with two or more determinants, each affected by the intervention or affecting the outcome but in no particular sequence. A workplace nutrition program provides an example of the multiple-determinant model (Kristal, Glanz, Tilley, & Li, 2000). The intervention featured at-work nutrition classes and self-help. The stakeholders and evaluators selected three determinants: predisposing factors (skills, knowledge, belief in diet-disease relationship); enabling factors (social support, perceived norms, availability of healthful foods); and stage of change (action and maintenance stages being subsequent to the intervention). The outcome variable was dietary change (eating vegetables and fruits). The model of this program is illustrated in Figure 10.6.

Kristal and colleagues found that the intervention did enhance predisposing factors as well as the likelihood of entering and remaining in the subsequent

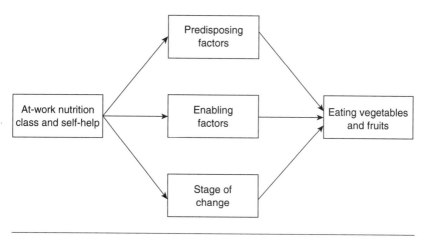

Figure 10.6   Workplace Nutrition Program as a Multiple Determinant, No
                   Sequential Order Model

SOURCE: Adapted from Kristal et al. (2000).

stages of change. They also found that the intervention did not affect enabling
factors. The program was failing because the intervention was failing to activate
one of the three determinants.

Multiple-Determinant Model With Sequential Order. The model containing
two or more determinants aligned in a causal order is a multiple-determinant
model with a sequential order. That is, certain determinants affect others in a
particular sequential order. An example of this kind of linear model is found
in an evaluation of a school-based antismoking campaign (Chen et al., 1988).
The intervention contained components such as an antismoking comic book,
discussions of the health messages the comic book delivered, and parental
notification about the intervention program. The determinants of the model, in
sequence, were the number of times the comic book was read, and knowledge
of the comic book's story and characters. The sequential order indicates that
repeated reading of the comic book changed the extent of knowledge about the
plot and characters. The sequence is illustrated in Figure 10.7.

The outcome to be measured was change in attitudes, beliefs, and behaviors related to smoking. The evaluation determined that the distribution of the
comic book affected the number of times the comic book was read, which in
turn affected knowledge of its content. However, neither of these determinants
was shown to affect students' smoking-related attitudes, beliefs, or behaviors.

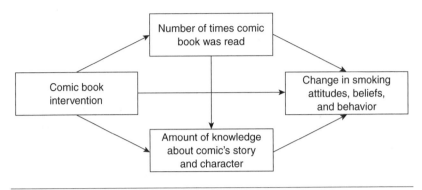

Figure 10.7   Antismoking Program as a Multiple Determinant With Sequential
Order Model

SOURCE: Adapted from Chen et al. (1988).

The Dynamic Model

The dynamic model of intervening mechanism evaluation assumes that multidirectional, reciprocal causal relationships exist among intervention, determinant, and outcome. The relationship between determinant and outcome, especially, is reciprocal rather than one-way: The determinant affects the outcome, and the outcome also affects the determinant. A hypothetical educational program illustrates the model well. The project's focus was to equip parents with skills and strategies to assist their children with homework; homework had been chosen as a determinant of primary students' school performance. The model made clear, however, that the relationship between parental involvement and student performance need not be linear. Parents becoming more involved in a child's schoolwork might improve the child's performance, and then seeing the improved performance, parents perhaps might feel gratified, stimulating their willingness to devote time and effort to remaining involved in the child's education. This form of the dynamic model is represented in Figure 10.8.

The dynamic model appears to be a sensible approach for many evaluation situations, but it is not widely applied at present by program evaluators. (The literature does, however, widely discuss one case of intervening mechanism evaluation using the dynamic model—an evaluation of the Transitional Aid Research Project by Berk, Lenihan, & Rossi, 1980. This study has several implications for theory-driven evaluation, including those noted in Shadish et al., 2002, and Chen, 1990.)

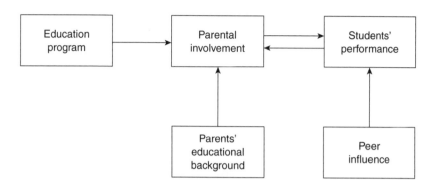

Figure 10.8   Education Program as a Dynamic Model

The reasons behind the limited use of the dynamic model to date include the high challenge of constructing a model that can be assessed with data actually available. Data analysis required in dynamic-model intervening mechanism evaluation relies on rather advanced statistical models such as the simultaneous equations model. These technical challenges aside, the dynamic model is a promising topic for future studies.

Some Theoretical Bases of Intervening Mechanism Evaluation

Research Methods Associated
With Intervening Mechanism Evaluation

Conducting an intervening mechanism evaluation usually requires both qualitative and quantitative research methods. In general, application of the intervening mechanism evaluation approach happens in two stages. In the model formulation stage, qualitative methods are essential for making explicit the stakeholders' change model. Methods such as the interview and the focus group are well suited to this task of the model formulation stage (see Chapter 4). In the data collection and analysis stage, quantitative methods become useful. They have a long tradition of collecting the data needed to test models and analyzing that data. For example, path analysis and structural equation modeling are well-established statistical techniques for testing models. The established nature of quantitative research methods is perhaps why most evaluators are given to applying these particular kinds of tests during an intervening mechanism evaluation. Note that this does not mean that qualitative methods cannot be useful to the evaluator here; but the question of the

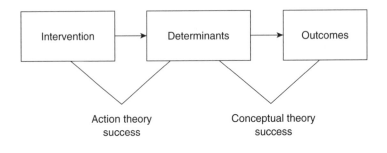

Figure 10.9   Action Theory and Conceptual Theory in the Intervening Mechanism
Evaluation

best ways to apply qualitative methods in testing change models is a topic for
further investigation.

When to Use an Intervening Mechanism Evaluation Approach

Thus far in the history of program development, intervention mechanism
evaluation approaches have tended to be applied when there is a need to assess
either the accuracy of causal chains or the relative efficacy of intervention
components.

Assessing Whether Causal Chains Functioned as Expected. Evaluators can
employ intervening mechanism evaluation to see which portion(s) of the
causal chains worked as projected and how this has contributed to the success
or failure of the program. Graphic representation of this kind of assessment is
provided in Figure 10.9.

The figure indicates that an intervening mechanism evaluation sets about
testing two theories. The first is the action theory, which is concerned with the
intervention's power to affect the determinant; the second is the conceptual
theory, which is concerned with the determinant's power to affect the outcome. The actual impact of causal chains is dependent on the success of both
the action theory *and* the conceptual theory. In the Adolescent Alcohol
Prevention Trial (Donaldson et al., 1994), which is cited above and illustrated
in Figure 10.4, we find an example of an intervention with accurate causal
chains. But even when poor action theory or poor conceptual theory is
revealed through an intervening mechanism evaluation, stakeholders should
benefit from the useful information that the evaluation obtains. Intervening
mechanism evaluation is useful in situations such as the following:

• *When Action Theory Fails.* Failed action theory means that an intervention fails to affect its determinant. When this occurs, but the conceptual program theory is sound, then at least the basic conceptualizing of the determinant as a cause of the problem is headed in roughly the right direction. Nevertheless, program designers need to restructure the program intervention to better activate the determinant.

• *When Conceptual Theory Fails.* Failed conceptual theory means that, although an intervention successfully activates its determinant, the determinant then fails to affect the outcome. This suggests that program designers have made an invalid assumption about causes of the problem—a fundamental flaw in their model that can be corrected by developing a better conceptual program theory. The above example of the antismoking program evaluated by Chen et al. (1988) illustrates conceptual program theory failure.

• *When the Two Theories Both Fail.* Data from an intervening mechanism evaluation indicating scant association between intervention and determinant, and furthermore between determinant and outcome, suggest that both action theory and conceptual theory are invalid. Continuing a program whose action theory and conceptual theory are both so flawed is usually not worthwhile. Heeding the lessons learned from the failure, program designers can begin at the beginning, with a reconceptualization of the problem, the intervention, and the determinant.

Assessing the Effectiveness of Intervention Components. Another venue for the application of intervening mechanism evaluation is the assessment of the relative effectiveness of an intervention's components. That is, for programs with multiple determinants, a case can be made for which of the determinants are most effective and which should be strengthened. Schneider, Ituarte, and Stokols (1993) evaluated a program promoting bicycle helmets. The program provided a "bicycle rodeo," physician education, direct-mail communication, coupons for discounted helmet purchase, and telephone communication. Using intervening mechanism evaluation, the researchers determined that the degree of parents' worry over bicycle accidents is a determinant mediating between the intervention and helmet ownership, an outcome. In addition, analysis of the relationship between intervention components and determinants identified the two intervention components having the greatest influence on children's helmet ownership. They were physician advising of need for a helmet and telephone communication—suggesting that rates of helmet use in a community are most likely to rise when an intervention employs interpersonal education efforts. Another study by Kristal et al. (2000), which has been employed earlier in the book, comprises a second example of applying intervening mechanism evaluation to compare components' effectiveness. The research team and stakeholders in a nutrition program had selected three determinants of the outcome, dietary change: predisposing factors, enabling factors, and stage of change. According to data, all three substantially affected dietary change, but only two—predisposing factors and stage of change—had, in fact, been affected by the intervention. The enabling factors determinant, the one not responding to the intervention, was proposed by the researchers to be alterable by some as-yet-unknown intervention component, resulting in an intensification of the overall intervention effect. In other words, to strengthen this program for future use, activity affecting the enabling factors should be explored, devised, and implemented.

THE MODERATING MECHANISM EVALUATION APPROACH

The second type of theory-driven outcome evaluation is the moderating mechanism evaluation. The moderating mechanism evaluation approach to theory-driven evaluation involves assessing one or more factors in a program's implementation that condition, or moderate, the intervention's effect on outcome. The factors are called *moderators*. In Figure 10.10, the basic model for the moderating mechanism evaluation is presented.

In Figure 10.10, the moderating mechanism is represented by the arrow drawn from each moderator to the midpoint of another arrow that is located between intervention and outcome, delineating the way in which the moderator conditions the intervention-outcome relationship. For example, the effectiveness of family counseling may depend on the trust maintained between counselor and clients. The level of trust maintained would be a moderator that conditioned the relationship between counseling and the outcome of counseling. Generally speaking, moderators can be clients' sociodemographic characteristics (e.g., race, gender, education, age); implementers' characteristics and styles (e.g., enthusiasm, commitment, skills, race, gender); features of client-implementer relationship (e.g., trust, compatibility, race, gender); the level of implementation fidelity; and the mode and setting of service delivery (e.g., formal vs. informal, centralized vs. decentralized, rural vs. urban, kind of organizational climate, and integrity of the intervention).

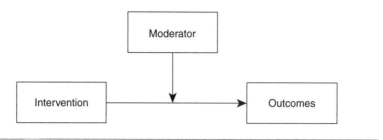

Figure 10.10   The Basic Model of Moderating Mechanism Evaluation

Constructing Moderating Mechanism Evaluation Models

The base of information needed to model a moderating mechanism evaluation can be drawn from stakeholders' ideas and experiences or from the literature. From there on, strategies used to construct the moderating mechanism model are those discussed above for construction of the intervening mechanism model. Up-to-date moderating mechanism evaluation mainly employs quantitative methods to construct models and analyze data. The equation used for analysis typically includes main effects (for intervention and moderator separately) and an interaction effect (intervention and moderator jointly). If the intervention is conditioned by the moderator, the interaction effect should be statistically significant.

Examples of Moderating Mechanism Evaluation

Two examples have been selected to demonstrate moderating mechanism evaluation: a case in which sociodemographic variables serve as moderators, and a case in which the integrity of the implementation serves as moderator. In the first example, tests were run on a sociodemographic variable, asking if it conditioned the relationships assumed within the model of an alcohol abuse prevention program (O'Leary, Jemmott, Goodhart, & Gebelt, 1996). The model is illustrated in Figure 10.11. The intervention had consisted of workshops in college classrooms and dormitories, as well as exhibits and a print (newspaper) campaign. Outcome measures were the number of sex partners and the total number of unprotected acts of sexual intercourse. The moderator was gender.

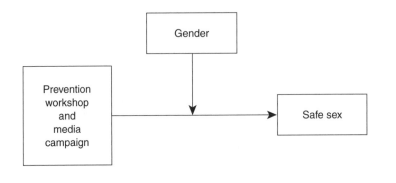

Figure 10.11   Social and Demographic Variables as Moderators
SOURCE: Adapted from O'Leary et al. (1996).

Analysis of the data collected concerning this program showed no statistically significant relationship between intervention and outcomes, which, however, was *not* tantamount to finding that no relationship between the two is possible. Indeed, further analysis of moderating processes in the model showed the intervention-outcome relationship to be moderated by gender. When all results were in, the evaluation concluded that the intervention had, in fact, reduced risky encounters among males substantially. It was among females that risky behavior remained unaffected.

In the second example of moderating mechanism evaluation, a valuable demonstration of program implementation's power to moderate the intervention-outcome relationship occurs. This evaluation tested whether quality of service delivery moderated any other component (Hansen, Graham, Wolkenstein, & Rohrbach, 1991). The intervention involved consisted of curricula for training youngsters to resist overt social pressure to use alcohol and other illegal substances, as well as for establishing conservative normative beliefs about the social and health consequences of substance use. The outcomes to be evaluated were seven indicators that pertained either to knowledge and perceptions of social pressure to use substances or to related refusal skills. Measures for quality of service delivery were ratings assigned by the program's expert stakeholders, along with a trained observer's ratings on items including enthusiasm, student responsiveness, student participation, classroom control, interaction with students, and meeting program goals. Figure 10.12 presents the model of this fairly complex program.

The evaluation suggested that intervention activities had affected some outcome measures. As importantly, it showed that the quality of service

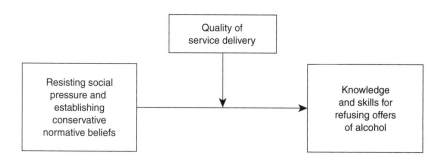

Figure 10.12   Service Delivery as a Moderator

SOURCE: Adapted from Hensen et al. (1991).

delivery had significantly moderated relationships between the intervention and three of the seven anticipated outcomes. These results led to a conclusion that the intervention worked best with high-quality service delivery.

Advanced Moderating Mechanism Models

In some cases, a moderating mechanism and an intervening mechanism are at work simultaneously in a program. Donaldson (2001) indicated that it is possible a moderator may condition between intervention and determinant or between determinant and outcome. The first possibility is illustrated in Figure 10.13.

The model in this figure requires that the effect of intervention on determinant be conditioned by a moderator, as in the earlier example of the "AIDS 101" intervention. In that example, the determinant supporting the intervention was increased HIV prevention knowledge and skills. Figure 10.14 contains a further example. It shows how a moderator is perhaps involved in the success of "AIDS 101"—for instance, the clients' education levels. An education moderator may quash the usefulness of "AIDS 101" with highly educated clients, possibly because their existing knowledge and skills are sufficient for avoiding HIV. On the other hand, moderation by education level may allow "AIDS 101" to be very useful with less educated clients, as they may well encounter little medical information in their daily lives.

The second possibility that a moderator conditions the relationship between determinant and outcome (Donaldson, 2001) is illustrated in Figure 10.14.

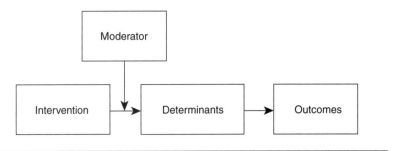

Figure 10.13   Moderating Mechanism Conditions the Relationship Between
Intervention and Determinant

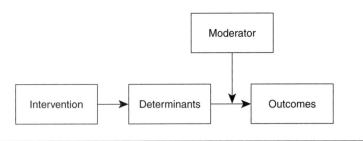

Figure 10.14   Moderating Mechanism Conditions the Relationship Between
Determinant and Outcome

Figure 10.14 diagrams the determinant's effect on outcome when this determinant is conditioned by a moderator. For a condom distribution program, the determinant might be availability of a condom. The model in Figure 10.14 demonstrates how condom availability's power to promote safe sex may be conditioned on sexual partners' acquiescence to values endorsing male dominance. Thus, even with condoms readily available, partners (especially men) who assume men's preferences should predominate are less likely than other individuals to use a condom for safe sex. On the other hand, the model suggests, partners valuing gender equality will perhaps use condoms if they are available.

In addition, there is another possibility: A moderator may condition at once the relationship of intervention to determinant and the relationship of determinant to outcome. The case is illustrated in Figure 10.15.

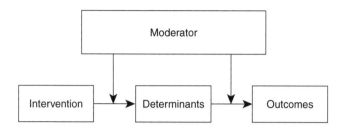

Figure 10.15 Moderating Mechanism Conditions Relationships Among
Intervention, Determinant, and Outcome

Dual conditioning of this type would be noted in a condom distribution program in which valuing of male dominance moderated partners' willingness to receive and carry condoms, as well as moderating their actual condom-using behavior.

When to Use a Moderating Mechanism Evaluation Approach

The moderating mechanism evaluation approach may be utilized when the purpose of evaluation is (a) identifying a need to tailor interventions for different groups, or (b) comparing the relative effectiveness of a program's structural options. Tailoring of interventions belies the widespread assumption that an intervention works or does not work for all kinds of target group members equally. The truth is that a target group's membership often comprises a variety of cultural, social, and economic backgrounds. This seems to imply that interventions should indeed be *expected* to work better for some members than others. The O'Leary research team's (1996) evaluation of the alcohol abuse/safe sex campaign showed that a moderating mechanism evaluation is equipped to comment on an intervention's generalizability across a target population. Moderating mechanism evaluation, in other words, discloses which intervention works for whom. Should strong differential effects of the intervention be deemed an issue within a target group, it becomes important to tailor versions of the intervention to the differential needs of subgroups within it.

As for the second task with which moderating mechanism evaluation can be charged, we must recognize the consistent truth that the structuring of a program could often be accomplished by various means. When more than one option presents itself, a moderating mechanism evaluation allows evaluators to formally test whether one option will be more effective than another. For example, stakeholders may wonder whether a particular intervention is especially effective if clients and implementers share a racial/ethnic background. Or, stakeholders may ask whether an intervention's efficacy depends on the delivery setting (e.g., home vs. workplace) or delivery mode (group vs. individual), and so on. A moderating mechanism evaluation approach is well-suited for assessing interaction between the intervention and other program elements.

THE INTEGRATIVE PROCESS/OUTCOME EVALUATION APPROACH

The third and final theory-driven outcome evaluation included in the book is the integrative process/outcome evaluation. This type of evaluation involves the systematic assessment of (a) the crucial assumptions beneath implementation, and (b) the causal processes of a program. In other words, integrative process/outcome evaluation weighs the range of aspects of a program theory (see Figure 10.2). This consummately comprehensive assessment provides a network of information about what works and what does not work in a program, from implementation processes to causal processes to effects on outcomes. Such a thorough analysis of potential pathways enlightens stakeholders as to how their program truly operates, providing the knowledge they will need to meet the accountability and program improvement requirements they face. Because this kind of comprehensive evaluation integrates both process and outcome evaluation, its relationship with other parts of a program can be illustrated as in Figure 10.16.

As indicated in Figure 10.16, an integrative process/outcome evaluation systematically assesses the implementation of the program plan and the truthfulness of the program rational.

One sees from the figure what a challenge it can be to design a successful intervention program. Creating a successful implementation of that program design in the field is also complex. The model makes clear that "implementation success" occurs only when an intervention appropriately activates a change process. Implementation success, then, is vital to the entire change process: If implementation fails, everything fails. Nevertheless, a successful implementation *guarantees* nothing about program success. If a program is to

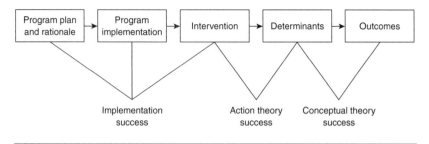

Figure 10.16   Linkages of Major Components in an Integrative Process/Outcome
Evaluation

be effective, its action theory and its conceptual program theory must succeed along with its implementation. Invalidity of either action theory or conceptual program theory could spell its doom. Comprehensive, systematic integrative process/outcome evaluation abundantly fleshes out assumed underlying mechanisms like these two theories comprise. It thus provides to stakeholders virtually any information they require for improving their programs.

Research Methods and Strategies Associated With Integrative Process/Outcome Evaluation

Integrative process/outcome evaluation often relies on mixed methods of collecting qualitative and quantitative data. Qualitative methods are preferred for formulating a model and acquiring data to test it, especially if the data describe some aspect of the implementation process. Designing an integrative process/outcome evaluation from start to finish involves the following general steps:

1. *Clarify the program theory.* Evaluators work with key stakeholders, including program designers, to clarify the program theory. This step incorporates both specifying the change model, or program rationale (see Chapter 4), and specifying the action model, or program plan (see Chapter 5).

2. *Collect and analyze data.* Guided by the program theory, empirical data are collected that will demonstrate how the theory operates in the field. Evaluators usually need mixed methods (qualitative and quantitative) to collect the necessary data pertinent to both action model and change model. Qualitative methods, however, are particularly useful in probing for the

reasons why a component is not working as well as expected. Data concerning linkages among program elements, from implementation to causal processes to final outcomes, need to be analyzed.

3. *Characterize the program in its entirety, then by its parts.* When all data have been analyzed and compiled, the evaluator provides to stakeholders the written overall assessment of whether the program as a whole is effectively reaching its goals. Incorporated in the report should be detailed analyses of the important parts of the program, as well, covering how well each is working and how each contributes to or hinders the achievement of program goals.

Examples of Integrative Process/Outcome Evaluation

Some readers may recognize that the garbage reduction program discussed in Chapter 2 is an example of integrative process/outcome evaluation. This section will turn to two other examples to illustrate the benefits of integrative process/outcome evaluation and to model some of the strategies involved.

Fort Bragg Child and Adolescent Mental Health Demonstration

The first example is a program offered by mental health services staff at the Army's Ft. Bragg, in North Carolina, which published its results as the Fort Bragg Child and Adolescent Mental Health Demonstration (Bickman, 1996). It was designed to assess whether the continuum of services the program provided actually improved treatment outcomes and reduced costs of care per client, as program goals required. The program provided traditional mental health services, such as outpatient therapy and acute inpatient care, and also more innovative services, such as case management, in-home therapy, after-school group treatment services, therapeutic homes, and 24-hour crisis management teams. Ft. Bragg families who requested services underwent comprehensive intake and assessment. In an attempt to control costs, the staff offered a continuum of services, at least some of which were expected to be appropriate for any given child. Evaluators used interviews, document review, and focus group meetings to develop the program theory for the project. Their theory is delineated in Figure 10.17.

The action model seen in the figure shows that the Fort Bragg Child and Adolescent Mental Health Demonstration emphasized the treatment and

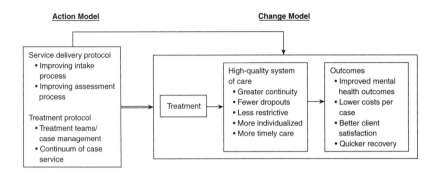

Figure 10.17 Program Theory of the Child and Adult Mental Health Demonstration

SOURCE: Adapted from Bickman (1996).

service-delivery protocols as well as the target group. The change model posits as determinant a better system of care, mediating between intervention and outcome. Evaluation data obtained suggested the Ft. Bragg demonstration had improved both access to care and quality of care, but clinical outcomes were disappointing. Little difference appeared between the demonstration program and comparison programs when it came to improved mental health, lower costs per client, quicker recovery, and client satisfaction. Thus, the demonstration program had implementation success and action theory success but a failure of its conceptual program theory.

Learnfare

A second example of integrative process/outcome evaluation is recorded in Ethridge and Percy's (1993) evaluation of a Wisconsin welfare program called Learnfare. Learnfare was, at heart, a welfare policy tying AFDC (Aid to Families with Dependent Children) payments to the school attendance of recipients' children. More specifically, children in families receiving AFDC payments were monitored by Learnfare to see how regularly they attended school. Learnfare sanctioned the family of a child who recorded two or more unexcused absences in a given month by cutting its monthly AFDC payment by the portion normally earmarked for that dependent child. The aim was to increase parents' involvement in schooling of their children, especially in terms of enhancing the educational achievement of teenagers, in an effort to decrease the children's chances of depending on AFDC in the future. Ethridge and Percy used program theory to understand how the policy was being

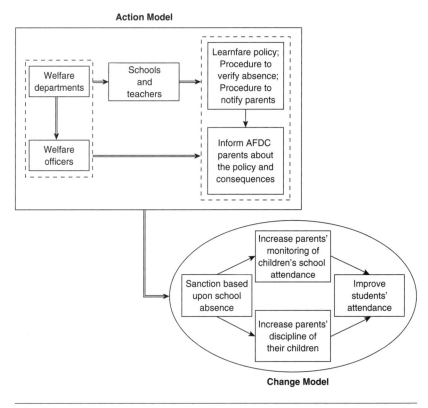

Figure 10.18   Program Theory Underlying Learnfare

SOURCE: Adapted from Ethridge and Percy (1993).

implemented and the limitation of the policy. The program theory underlying Learnfare is illustrated in Figure 10.18.

Figure 10.17 indicates that for the portion of the change model that underlay the Learnfare policy, Ethridge and Percy identified two determinants that were crucial to the policy's effectiveness: (a) an increase in parents' responsiveness to attendance-related communications from schools, combined with an increase in parents' own monitoring of children's school attendance; and (b) increased capacity of parents to control school attendance-related behaviors of children. The evaluators questioned the validity of the first determinant. Low rates of literacy are common among parents receiving AFDC, they pointed out; it is not farfetched to propose that many such parents are unable to read and understand communications from schools concerning attendance. The authors also felt that there were problems with the second determinant stakeholders had specified. They suggested that because many families are single-parent households, impoverished and susceptible to various physical and social problems, expecting parents to routinely monitor and control children's behavior might be unrealistic. Thus, the change model on which Learnfare was founded could not, they believed, operate as stakeholders anticipated.

Learnfare's action model also exhibited major weaknesses, according to Ethridge and Percy (1993), especially in its target group, implementing organizations, and relationships with peer organizations. Some of the problems that the evaluators brought up seem unmistakable in hindsight:

- *"Unexcused absence" was not clearly defined.* Learnfare had given no comprehensive definition of an unexcused absence with which to explain its attendance requirements in the context of schools' attendance requirements and make the program workable. With unexcused absences triggering sanctions, a precise explanation of what unexcused absences are had to be given. Schools' individual policies about acceptable absences showed greatest variation when it came to family vacations, hunting or fishing trips, and family business activities.

- *Gathering attendance information was difficult.* Learnfare depended on the transmittal of attendance information from school to welfare system. This sounds straightforward until one realizes that state law mandates the privacy of school records. Unauthorized sharing of attendance data between schools and county or state welfare systems violated the law; so each AFDC parent needed to waive the right to privacy and direct school officials to reveal the data. This seriously delayed transmission of attendance data. Furthermore, for Learnfare to succeed, schools needed to transmit accurate data in a timely way, and this proved problematic. Schools had difficulty keeping current their attendance data on AFDC students. County welfare officials needed to act on data that often arrived late, was hard to interpret, and came from many idiosyncratic schools districts. Another law required schools to follow elaborate verification procedures to increase accuracy of attendance reports. In the wake of all this, 3 or 4 months could pass between the unexcused absence and the punitive loss of funds.

- *Communication with AFDC recipients was complicated.* If Learnfare were to work, AFDC recipients had to be aware of the sanctions and the reasons for them. They also needed to be notified when a child's truancy had triggered a benefit reduction. A substantial effort was made to disseminate general information about Learnfare to the target group. However, the problem of adequate notification of parents about specific dates of attendance violations persisted. It proved difficult for the program to state for parents the dates of first and second (and subsequent) unexcused absences. This was the result of an unproductive relationship among welfare officials and local school officials. School officials had not been invited to help plan for Learnfare and its attendant impacts on educational objectives, support personnel, a school's own attendance recording and monitoring procedures, and nearby alternative schools.

THEORY-DRIVEN OUTCOME
EVALUATION AND UNINTENDED EFFECTS

In evaluation parlance, unintended effects are the effects made by a program that are outside the scope of program goals. The tendency for social action to generate unintended effects has long been articulated in the literature (e.g., Weber, 1947). However, when enlightenment is a central purpose of evaluation, it is possible for the evaluator to manage the potential for unintended effects using his or her evaluation design. Unintended effects can be either positive or negative. Incarcerating juvenile delinquents perhaps has the negative effect of allowing them to learn criminal skills from veteran offenders. The 55-mile-per-hour speed limit, in contrast, was intended to create a fuel conservation effect, then was also found to boost highway safety, a positive effect (Clotfelter & Hahn, 1978). A great advantage of theory-driven outcome evaluation is the detection of unintended effects that it facilitates. Unintended effects are detected during theory-driven outcome evaluation by two general strategies: the use of formal specification and the use of field study.

Formal Specification of Possible Unintended Effects

A strategy to specify formally any possible unintended effects involves rendering each such effect as a hypothesis that the evaluation can test. The evaluator has two options here: to use existing work to infer unintended effects or to ask stakeholders to brainstorm possible unintended effects. The former option tasks evaluators with reviewing existing theories and studies that are relevant to the program. This permits the formulation of hypotheses about possible unintended effects of the current program, which can then be tested. For example, in evaluating compulsory seatbelt legislation and its effect on vehicle occupant casualties, Conybeare (1980) harnessed existing theory and knowledge and posited the potential unintended effect of a rise in the casualty rate for nonoccupants. Conybeare's reasoning was that drivers experiencing the legislation-induced security of being strapped in might begin to drive less attentively, resulting in more accidents with pedestrians, bicyclists, and so on. Such a hypothesis was included in the evaluation design. The data confirmed that the legislation had created both the stakeholders' intended and the evaluator's unintended effect.

The second option for detecting unintended effects during evaluation involves prompting stakeholders to brainstorm possible unintended effects. Implementers and other stakeholders offer a great deal of experience in working with clients, communities, and implementations. They comprise a good source of ideas about unintended effects worthy of inclusion in an evaluation. For example, I have participated in evaluating an after-school project designed to offer academic tutoring, recreation activities, and drug use prevention education to adolescents of South Asian ethnic background. The program goals were to enhance the target group's success in school and cultivate in them attitudes and skills for resisting drug use. As the program theory was being formulated, counselors with the program pointed out an important potential unintended effect, which was that the interaction fostered by program activities might weaken the participants' prejudices toward Asian ethnic groups not their own. In many Asian communities, including the one that was home to this project, interethnic group prejudices are a pressing concern, as they often lead to conflict and even violence. The hypothesis suggested by the program counselors was included in the evaluation design.

Field Study Detection of Unintended Effects of Implementation

The integrative process/outcome evaluation usually combines both qualitative and quantitative methods in the evaluation in order to intensively examine an implementation process. The evaluator in the field will benefit from understanding how a program is being implemented, how its clients and implementers are responding to it, and how it is interacting with its environment. Field observation provides a good opportunity for evaluators to identify and investigate potential unintended effects. For example, I also had reason to visit the site where a successful alcohol treatment program was being implemented within an institution that confined its residents. The quality of implementation was very impressive. But it was also clear that many clients smoked cigarettes during breaks and after meals. Perhaps smoking relieved the stress of eschewing alcohol. Staff acknowledged that smoking by clients was permitted. This field observation led to a hypothesis about an unintended effect of the program: Clients who cease consuming alcohol may begin to consume tobacco products.

The assessment by Gottlieb et al. (1992) of the effects of a workplace smoking policy (used as an example in Chapter 7) in addition provides a good illustration of the detection of unintended effects during implementation. In interviewing smokers and nonsmokers employed by the affected organization, the study found four unintended effects:

- *Diminished Air Quality in Designated Smoking Areas.* Smokers were relegated to certain smoking areas, where they tended to congregate, filling the air with the smoke of many cigarettes rather than one. Air quality was considerably diminished in the smoking areas, especially poorly ventilated ones. The odor of cigarette smoke became so thick that it reached even to nonsmokers who were at a distance from the smoking areas.
- *Disruption of Work.* Implementing the smoking policy led to the disruption of work. Clerical staff left their desks more often; nonclerical employees began to move their work with them to smoking areas and the cafeteria.
- *Disruption of Communication.* Once the policy was in place, nonsmokers were less likely than before to discuss smoking-related issues with smokers; the percentage of nonsmokers requesting a coworker not to smoke near them decreased. Some employees expressed concern that the smoking policy polarized the smoking and nonsmoking employees.
- *Purported Creation of "Deviance."* After the policy had been implemented, employees surfaced who said they were concerned that tobacco use was being demonized, resulting in smokers being discriminated against more noticeably. Smokers were, they said, increasingly viewed as deviants.

The use of both quantitative and qualitative methods during integrative process/outcome evaluation opens an opportunity for evaluators to appraise a program's unintended effects. In the Chapter 7 example of the anti-drug program launched by the Taiwanese Ministry of Education (Chen, 1997), quantitative outcome data argued that the program was a strong success. However, some substantive problems identified in the implementation process cast doubt on that conclusion, in spite of the quantitative data. Evaluators returned to the field to interview key informants and probe them about these problems. They learned that in the local schools staff feared the Ministry of Education would use the data to punish administrators and teachers at schools with more drug-using students. They sought to protect themselves by reporting fewer cases than they knew to exist. The outcome data they provided were not a valid reflection of the program's effect.

⊰ ELEVEN ⊱

LOOKING FORWARD

I t is commonly understood that it is undesirable to conduct a one-size-fits-all
kind of evaluation. But the real challenge is where and how to go from here.
Now, after reading this book, the readers can take on the challenge and are capable of answering the following questions: How do we figure out stakeholders'
evaluation needs and evaluation circumstances? How many practical evaluation
strategies and approaches are available to apply? How many evaluation options
are available to deal with a particular evaluation need and circumstance? How
should we select an appropriate evaluation option? How should we actually
conduct the evaluations? To gain and possess knowledge in these areas is quite
an accomplishment.

This book enriches evaluators' toolbox and provides them with the
following tools:

- Practical evaluation taxonomy is developed to help evaluators and
  stakeholders conceptualize evaluation needs and circumstances, as
  well as available evaluation strategies and approaches. The road map
  helps evaluators and stakeholders to systematically and proactively
  plan current evaluation in a stage and future evaluations in other stages.
- Evaluation options that match stakeholders' evaluation needs are identified. The trade-offs among these options are discussed to help evaluators and stakeholders make an informed decision when selecting an
  evaluation approach or approaches for their program.
- The procedures and methods for conducting each evaluation approach
  are discussed and illustrated with concrete practical examples.

- Evaluation strategies and approaches in major program stages (program planning, initial implementation, mature implementation, and outcome) are systematically and meaningfully linked under the conceptual framework of program theory.
- The conceptual framework of program theory and related concepts are systematically introduced. Their relevance and usefulness for working with stakeholders to conceptualize a program and identify potential solutions are illustrated.

By addressing these issues, the book could contribute to the future advancement of program evaluation in the following areas.

EVALUATORS' PLURAL ROLES

In the early history of program evaluation, some evaluators had portrayed the evaluators' role as that of "objective assessors." With the progress of program evaluation, the importance and necessity of participatory evaluation has been realized. We found out that we are rarely playing a pure role of objective assessor. Often, we play an assessor role with a need to interact with stakeholders and consider their views and concerns in designing an evaluation. Besides the role of assessor, many other roles are identified in the evolution of program evaluation. This book discusses a variety of evaluators' roles to play depending on which program stage and evaluation needs are focused and which evaluation strategy and approach are applied. In addition to assessor, the other roles are background information providers, troubleshooters, facilitators, consultants, development partners, and illuminators. The meanings and implications of evaluators' plural roles have to be addressed more systematically in the future.

Chapter 3 discusses an issue of potential role conflict resulting from evaluators conducting evaluation across different stages. However, many additional issues need to be discussed. For example, with multiple evaluators' roles, how do we educate or train evaluation students in the future? Should we train students as specialists in one role? For example, perhaps some students take the facilitator role, others take that of the assessor or the troubleshooter. Or, should we train them as generalists? Some students have good knowledge of all these roles. Or, should we have options such as "major" in one role and "minor" in another role? In addition, we also need to study how the market reacts to these roles and what the professional standards are for maintaining or ensuring the quality of their work.

POLITICS OF PROGRAM EVALUATION

Program evaluation is conducted in a political arena. Evaluation is just one ingredient in the decision-making process. The evaluation is most useful when it contributes to the political process. In order to have such an achievement, at first stakeholders and evaluators need to have a mutual understanding on the evaluation needs and the evaluation approaches to be used, and have stakeholders participate in the evaluation process. The mutual understanding and agreement could reduce the likelihood of an evaluation being attacked or criticized by someone with a purely political motive. For example, without such understanding and agreement, stakeholders can easily dismiss the merit of the evaluation by stating that it did not evaluate what was being accomplished. Politics are more likely to happen where there are confusion and uncertainty. These issues on how to identify and reduce confusion and uncertainty in the evaluation process are important and need to be discussed in the future.

EVALUATION PRACTITIONERS' CONTRIBUTIONS TO THE DEVELOPMENT OF EVALUATION THEORIES AND METHODOLOGY

Because program evaluation is still relatively young as an applied science, the development of evaluation theories and methodology have mainly been using a top-down approach; that is, a small group of talented theorists and methodologists develops the body of evaluation theories and methodology for evaluation practitioners to apply. The approach has worked well in the past; evaluators are in need of basic concepts, strategies, and tools to follow to begin their work. However, the further advancement of program evaluation requires evaluation practitioners to actively participate in developing evaluation theories and methodologies. In short, a bilateral approach is needed to advance program evaluation. Evaluation practitioners must be part of the picture for the following reasons.

Providing Empirical Feedback for Refining the Existing Theories, Approaches, and Methodologies

Many evaluation theories, approaches, and methodologies exist in literature. It is high time for evaluation practitioners to provide their experience in applying them. They can provide valuable information on which parts of these theories, approaches, and methodologies are working and which parts need to be revised or changed, and how to do it.

The body of evaluation knowledge needs empirical feedback to nurture its growth.

Pointing Out the Areas That Need to Develop and Expand

Because evaluation practitioners are working in the field every day, they know which evaluation areas to expand, which new areas need to develop, and which areas have become obsolete. Without their constant feedback for corrections, evaluation might move in a direction that indulges in abstract ideas and debates. We need practitioners' participation for enhancing the practical relevance of the body of evaluation knowledge.

Future efforts would require increasing the use of the bilateral approach for advancing the body of evaluation knowledge; perhaps this book could be a first step in that direction.

NARROWING THE GAPS BETWEEN ACADEMIC AND PRACTICAL COMMUNITIES

The advancement of social and health promotion interventions requires close interactions, communication, and collaborations between academic and practical communities. In the collaboration, on one hand, intervention scientists produced highly promising and feasible interventions for practitioners to use. On the other, intervention practitioners apply these evidence-based and practice-oriented interventions in their day-to-day service delivery, and they give feedback on their experience of such a practice and their future needs. The synergy resulting from such collaboration benefits each other's growth.

Unfortunately, this does not happen in many social and health promotion interventions. There is a big gap between academic and practical communities.

Using HIV prevention as an example, many innovative interventions have been studied and published in the literature. However, fewer actual applications of these evidence-based interventions by CBOs (who are foot soldiers of HIV prevention) have been studied. CBOs are practicing the conventional interventions with which they are familiar and comfortable.

According to the bilateral paradigm discussed in Chapter 9, improving the problem requires a bridge for bringing both communities together simultaneously. Many current efforts—such as the call for doing more effectiveness evaluations and translation research, and providing more capacity building and technical assistance to practitioners and their organizations for enhancing dissemination—are encouraging. These efforts will help to bring scientists closer to the field. However, there has been much less talk about how to understand the practitioners' view and practice and how to prepare and motivate them to collaborate with the scientific community. I have frequently heard practitioners' complaints that many evaluation funds are used to evaluate innovative programs, and few funds are available for them to evaluate their existing programs. As a result, they and scientists are not able to benefit from learning what works and what does not in current practice. Their views and concerns are informative. Perhaps the scientific community and funding agencies should rethink their priority and put more research efforts into understanding stakeholders' program theory and conducting evaluation to help stakeholders improve their programs. Perhaps when these efforts are committed to by funding agencies and academic community, practitioners will see the benefits of using more and more evidence-based interventions in their practice. This discussion of program theory can facilitate bridging the gap between academic and practical communities.

REFERENCES

————•••————

Ajzen, I., & Fishbein, M. (Eds.). (1980). *Understanding attitudes and predicting social behavior.* Englewood Cliffs, NJ: Prentice Hall.

Alabama Comprehensive Tobacco Use Prevention and Control Plan. (2000). Birmingham: Alabama Department of Health.

Bandura, A. (1977). *Social learning theory.* Englewood Cliffs, NJ: Prentice Hall.

Barbacki, S., & Fisher, R. A. (1936). A test of the supposed precision of systematic arrangement. *Annals of Eugenics, 7,* 189-193.

Bartholomew, L. K., Parcel, G. S., Kok, G., & Gottlieb, N. H. (2001). *Intervention mapping: Designing theory- and evidence-based health promotion programs.* Mountain View, CA: Mayfield.

Berk, R. A., Lenihan, K. J., & Rossi, P. H. (1980). Crime and poverty: Some experimental evidence for ex-offenders. *American Sociological Review, 45,*766-786.

Bickman, L. (1987a). The function of program theory. In L. Bickman (Ed.), *Using program theory in evaluation.* San Francisco: Jossey-Bass.

Bickman, L. (1987b). *Using program theory in evaluation.* San Francisco: Jossey-Bass.

Bickman, L. (Ed.). (1990). *Advances in program theory.* San Francisco: Jossey-Bass.

Bickman, L. (1996). The application of program theory to a managed mental health care evaluation. *Evaluation and Program Planning, 19*(2), 111-119.

Blakely, C. H., Mayer, J. P., Gottschalk, R. G., Schmitt, N., Davidson, W. S., Roitman, D. B., & Emshoff, J. G. (1987). The fidelity-adaptation debate: Implications for the implementation of public sector social programs. *American Journal of Community Psychology, 15,* 253-268.

Boruch, R. F. (1977). *Randomized experiments for planning and evaluation: A practical guide.* Thousand Oaks, CA: Sage.

Brunk, S. E., & Goeppinger, J. (1990). Process evaluation: Assessing re-invention of community-based interventions. *Evaluation & the Health Professions, 13*(2), 186-203.

Centers for Disease Control & Prevention (CDC). (2001). Evaluating CDC-funded health department HIV prevention programs. Atlanta: CDC, National Centers for HIV, STD & TB Prevention.

Chen, H. T. (1990). *Theory-driven evaluations.* Newbury Park, CA: Sage.

Chen, H. T. (1994). Current trends and future directions in program evaluation. *Evaluation Practice, 15*(1), 79-82.

Chen, H. T. (1996). A comprehensive typology for program evaluation. *Evaluation Practice, 17*(2), 121-130.

Chen, H. T. (1997). Normative evaluation of an anti-drug abuse program. *Evaluation and Program Planning, 20*(2), 195-204.

Chen, H. T. (2001). Development of a national evaluation system to evaluate CDC-funded health department HIV prevention programs. *American Journal of Evaluation, 22*(1), 55-70.

Chen, H. T. (2003). Theory-driven approach for facilitation of planning health promotion or other programs. *Canadian Journal of Program Evaluation, 18*(2), 91-113.

Chen, H. T. (2004). The roots of theory-driven evaluation. In M. C. Alkin (Ed.), *Evaluation roots: Tracing theorists' views and influences.* Thousand Oaks, CA: Sage.

Chen, H. T., & Mark, M. (1996). Assessing the needs of inner city youth: Beyond needs identification and prioritization. *Children and Youth Services Review, 18*(3), 1-31.

Chen, H. T., Quane, J., & Garland, T. N. (1988). Evaluating an antismoking program. *Evaluation & the Health Professions, 11*(4), 441-464.

Chen, H. T., & Rossi, P. H. (1980). The multi-goal, theory-driven approach to evaluation: A model linking basic and applied social science. *Social Forces, 59,* 106-122.

Chen, H. T., & Rossi, P. H. (1992). *Using theory to improve program and policy evaluations.* Westwood, CT: Greenwood.

Chen, H. T., Wang, J. C. S., & Lin, L. H. (1997). Evaluating the process and outcome of a garbage reduction program in Taiwan. *Evaluation Review, 21*(1), 27-42.

Clapp, J. D., & Early, T. J. (1999). A qualitative exploratory study of substance abuse prevention outcomes in a heterogeneous prevention system. *Journal of Drug Education, 29*(3), 217-233.

Clark, G. N. (1995). Improving the transition from basic efficacy research to effectiveness studies: Methodological issues and procedures. *Journal of Consulting and Clinical Psychology, 63,* 718-725.

Clotfelter, C. T., & Hahn, J. F. (1978). Assessing the national 55 mph speed limit. *Policy Sciences, 9,* 281-294.

Connell, J. P., Kubisch, A. C., Schorr, L. B., & Weiss, C. H. (Eds.). (1995). *New approaches to evaluating community initiatives: Concepts, methods and contexts.* Washington, DC: Aspen Institute.

Conybeare, J. A. (1980). Evaluation of automobile safety regulations: The case of compulsory seat belt legislation in Australia. *Policy Sciences, 12,* 27-39.

Cook, T. D., & Campbell, D. T. (1979). Quasi-experimentation: Design & analysis issues for field settings. *Boston: Houghton Mifflin.*

Cordray, D. S. (1986). Quasi-experimental analysis: A mixture of methods and judgment. In W. M. K. Trochim (Ed.), *Advances in quasi-experimental design and analysis* (New Directions for Program Evaluation, No. 31). San Francisco: Jossey-Bass.

Cronbach, L. J. (1982). *Designing evaluations of educational and social programs.* San Francisco: Jossey-Bass.

Davis, M., Baranowski, T., Resnicow, K., Baranowski, J., Doyle, C., Smith, M., Wang, D. T., & Yaroch, A. (2000). Gimme 5 fruit and vegetables for fun and health: Process evaluation. *Health Education Behavior, 27*(2), 167-176.

Donaldson, S. I. (2001). Mediator and moderator analysis in program development. In S. Sussman (Ed.), *Handbook of program development for health behavior research & practice* (pp. 470-496). Thousand Oaks, CA: Sage.

Donaldson, S. (2003). Theory-driven program evaluation in the new millennium. In S. Donaldson & M. Scriven (Eds.), *Evaluating social programs and problems* (pp. 109-141). Mahwah, NJ: Lawrence Erlbaum.

Donaldson, S. I. , Graham, J. W., & Hansen, W. B. (1994). Testing the generalizability of intervening mechanism theories: Understanding the effects of adolescent drug abuse prevention interventions. *Journal of Behavioral Medicine, 17*(2), 195-215.

Durkheim, E. (1965). *The division of labor in society.* New York: Free Press.

Ethridge, M. E., & Percy, S. L. (1993). A new kind of public policy encounters disappointing results: Implementing Learnfare in Wisconsin. *Public Administration Review, 54*(4), 340-347.

Fetterman, D. M., Kaftarian, S. J., & Wandersman, A. (Eds.). (1996). *Empowerment evaluation: Knowledge and tools for self-assessment & accountability.* Thousand Oaks, CA: Sage.

Flay, B. R. (1986). Efficacy and effectiveness trials (and other phases of research) in the development of health promotion programs. *Preventive Medicine, 15,* 451-474.

Foster, G. H. (1973). *Traditional societies and technical change.* New York: Harper & Row.

Fulbright-Anderson, K., Kubisch, A. C., & Connell, J. P. (1998). *New approaches to evaluating community initiatives: Vol. 2: Theory, measurement and analysis.* Washington, DC: Aspen Institute.

Gettleman, L., & Winkleby, M. A. (2000). Using focus groups to develop a heart disease prevention program for ethnically diverse, low-income women. *Journal of Community Health, 25*(6), 439-453.

Glantz, K., Carbone, E., & Song, V. (1999). Formative research for developing targeted skin cancer prevention for children in multi-ethnic Hawaii. *Health Education Research: Theory and Practice, 14*(2), 155-166.

Glasgow, R. E., Lando, H., Hollis, J., McRae, S. G., & La Chance, P. A. (1993). A stop-smoking telephone help line that nobody called. *American Journal of Public Health, 83*(2), 252-253.

Glasgow, R. E., Lichtenstein, E., & Marcus, A. (2003). Why don't we see more translation of health promotion research to practice? Rethinking the efficacy-to-effectiveness transition. *American Journal of Public Health, 93*(8), 1261-1267.

Goodman, R. M., Wandersman, A., Chinman, M., Imm, P., & Morrissey, E. (1996). An ecological approach of community-based interventions for prevention and health promotion: Approaches to measuring community coalitions. *American Journal of Community Psychology, 24*(1), 33-61.

Gottlieb, N. H., Lovato, C. Y., Weinstein, R., Green, L. W., & Eriksen, M. P. (1992). The implementation of a restrictive worksite smoking policy in a large decentralized organization. *Health Education Quarterly, 19*(1), 77-100.

Gowdy, E. A., & Freeman, E. (1993). Program supervision: Facilitating staff participation in program analysis, planning and change. *Administration in Social Work, 17*(3), 59-79.

Green, L. W. (2001). From research to "best practices" in other settings and ppulations. *American Journal of Health Behavior, 25,* 165-178.

Green, L. W., & Kreuter, M. W. (1991). Health promotion planning: An educational and environmental approach. Mountain View, CA: Mayfield.

Greene, J., & Caracelli, V. (1997). *Advances in mixed-method evaluation: The challenges and benefits of integrating diverse paradigms* (New Directions for Evaluation, No. 74). San Francisco: Jossey-Bass.

Greenwald, P., & Cullen, J. W. (1985). The new emphasis in cancer control. *Journal of the National Cancer Institute, 74,* 543-551.

Guba, E. G., & Lincoln, Y. S. (1989). *Fourth-generation evaluation.* Newbury Park, CA: Sage.

Hansen, W. B., Graham, J. W., Wolkenstein, B. H., & Rohrbach, L. A. (1991). Program integrity as a moderator of prevention program effectiveness: Results for fifth-grade students in the Adolescent Alcohol Prevention Trial. *Journal of Studies on Alcohol, 52*(6), 568-579.

Hawe, P., & Stickney, E. K. (1997). Developing the effectiveness of an intersectoral food policy coalition through formative evaluation. *Health Education Research, 12*(2), 213-225.

Jones, J. H. (1981). *Bad blood: The Tuskegee syphilis experiment.* New York: Free Press.

Kohler, C. L., Grimley, D., & Reynolds, K. D. (1999). Theoretical approaches guiding the development and implementation of health promotion programs. In J. M. Raczynski & R. J. DiClemente (Eds.), *Handbook of health promotion and disease prevention* (pp. 23-50). New York: Kluwer Academic.

Kristal, A. R., Glanz, K., Tilley, B. C., & Li, S. (2000). Mediating factors in dietary change: Understanding the impact of a worksite nutrition intervention. *Health Education & Behavior, 27*(1), 112-125.

Krueger, R. A. (1988). *Focus groups: A practical guide for applied research.* Newbury Park, CA: Sage.

Laub, C., Somera, D. M., Gowen, L. K., & Diaz, R. M. (1999). Targeting "risky" gender ideologies: Constructing a community-driven theory-based HIV prevention intervention for youth. *Health Education & Behavior, 26*(2),185-199.

Mark, M. M. (2003). Toward an integrative view of the theory and practice of program and policy. In S. Donaldson & M. Scriven (Eds.), Evaluating social programs and problems (pp. 183-204). Mahwah, NJ: Lawrence Erlbaum.

Mark, M. M., Henry, G. T., & Julnes, G. (2000). *Evaluation: An integrated framework for understanding, guiding and improving policies and programs.* San Francisco: Jossey-Bass.

Mark, M. M., Hofmann, D. A., & Reichardt, C. S. (1992). Testing theories in theory-driven evaluation: (Test of) moderation in all things. In H. T. Chen & P. H. Rossi (Eds.), *Using theory to improve program and policy evaluations* (pp. 71-84). Westport, CT: Greenwood.

Marx, R., Hirozawa, A. M., Chu, P. L., Bolan, G. A., & Katz, M. H. (1999). Linking clients from HIV antibody counseling and testing to prevention services. *Journal of Community Health, 24*(3), 201-214.

McHugo, G. J., Drake, R. E., Teague, G. B., & Xie, H. (1999). Fidelity to assertive community treatment and client outcomes in the New Hampshire dual disorders study. *Psychiatric Services, 50*(6), 818-824.

McKillip, J. (1992). Research without control groups. In F. B. Bryant, J. Edwards, R. S. Tindale, E. J. Posavac, L. Heath, E. Henderson-King, & Y. Suarez-Balcazar (Eds.), *Methodological issues in applied social psychology.* New York: Plenum.

Mercier, C., Piat, M., Peladeau, N., & Dagenais, C. (2000). An application of theory driven evaluation to a drop-in youth center. *Evaluation Review, 24*(1),73-91.

Miller, W. R., Toscova, R. T., Miller, J. H., & Sanchez, V. (2000). A theory-based motivational approach for reducing alcohol/drug problems in college. *Health Education & Behavior, 27*(6), 744-759.

Oldenburg, B. F., Ffrench, M. L., & Sallis, J. F. (2000). Health behavior research: The quality of the evidence base. *American Journal of Health Promotion, 14,* 253-257.

O'Leary, A., Jemmott, L. S., Goodhart, F., & Gebelt, J. (1996). Effects of an institutional AIDS prevention intervention: Moderation by gender. *AIDS Education and Prevention, 8*(6), 516-528.

Patton, M. Q. (1997). *Utilization-focused evaluation: The new century text* (3rd ed.). Thousand Oaks, CA: Sage.

Paul, G. L. (1966). *Insight versus desensitization in psychotherapy.* Stanford, CA: Stanford University Press.

Pawson, R., & Tilly, N. (1997). *Realistic evaluation.* Thousand Oaks, CA: Sage.

Perry, P. D., & Backus, C. A. (1995). A different perspective on empowerment in evaluation: Benefits and risks to the evaluation process. *Evaluation Practice, 16*(1), 37-46.

Quantock, C., & Beeynon, J. (1997). Evaluating an osteoporosis service using a focus group. *Nursing Standard, 11*(42), 45-47.

Rogers, P. J., Hasci, T. A., Petrosino, A., & Huebner, T. A. (Eds.). (2000). *Program theory in evaluation: Challenges and opportunities* (New Directions for Evaluation, No. 87). San Francisco: Jossey-Bass.

Rossi, P., Lipsey, M. W., & Freeman, H. E. (2004). *Evaluation: A systematic approach* (7th ed.). Thousand Oaks, CA: Sage.

Schneider, M., Ituarte, P., & Stokols, D. (1993). Evaluation of a community bicycle helmet promotion campaign: What works and why. *American Journal of Health Promotion, 7*(4), 281-287.

Scriven, M. (1967). The methodology of evaluation. In R. E. Stake et al. (Eds.), *Perspectives on curriculum evaluation* (AERA Monograph Series on Curriculum Evaluation, No. 1). Chicago: Rand McNally.

Scriven, M. (1998). Minimalist theory: The least theory that practice requires. *American Journal of Evaluation, 19*(1), 57-78.

Shadish, W. T., Cook, T. D., & Campbell, D. T. (2002). *Experimental and quasi-experimental designs for generalized causal inference.* Boston: Houghton Mifflin.

Shadish, W. R., Cook, T. D., & Leviton, L. C. (1991). *Foundations of program evaluation: Theories of practice.* Newbury Park, CA: Sage.

Shapiro, J. P., Secor, C., & Butchart, A. (1983). Illuminative evaluation. *Educational Evaluation and Policy Analysis, 5*(4), 465-471.

Sloane, R. B., Staples, F. R., Cristol, A. H., Yorkston, N. J., & Whipple, K. (1975). *Psychotherapy versus behavior therapy.* Cambridge, MA: Harvard University Press.

Smith, M. F. (1989). *Evaluability assessment: A practical approach.* Norwell, MI: Kluwer.

Strecher, V. J., Bauman, K. E., Boat, B., Fowler, M. G., Greenberg, R. A., & Stedman, H. (1989). The development and formative evaluation of a home based intervention to reduce passive smoking by infants. *Health Education Research: Theory and Practice, 4*(2), 225-232.

Strecher, V. J., & Rosenstock, I. M. (1997). The health belief model. In K. Glanz (Ed.), *Health behavior and health education: Theory, research, and practice.* San Francisco: Jossey-Bass.

Student. (1936a). Co-operation in large scale experiments. *Journal of the Royal Statistical Society (Series B), 3,* 115-136.

Student. (1936b). The half-drill system agricultural experiments. *Nature, 138,* 971-972.

Stufflebeam, D. L. (Ed.). (2001). *Evaluation models* (New Directions for Evaluation, No. 89). San Francisco: Jossey-Bass.

Suchman, E. (1967). *Evaluation research.* New York: Russell Sage.

Suchman, E. (1969). Evaluating educational programs. *Urban Review, 3*(4), 15-17.

Trochim, W. M. K. (1984). *Research design for program evaluation: The regression discontinuity approach.* Beverly Hills, CA: Sage.

Trochim, W. M. K. (1998). An evaluation of Michael Scriven's *Minimalist theory: The least theory that practice requires. American Journal of Evaluation, 19*(2), 243-249.

Trochim, W. M. K., & Cook, J. A. (1992). Pattern matching in theory-driven evaluation: A field example from psychiatric rehabilitation. In H. T. Chen & P. H. Rossi (Eds.), *Using theory to improve program and policy evaluations* (pp. 49-69). Westport, CT: Greenwood.

United Way of America Task Force on Impact. (1996). *Measuring outcome: A practical approach.* Alexandria, VA: United Way of America.

Wallin, E., Bremberg, S., Haglund, B., & Holm, L. E. (1993). Cancer prevention in schools: Design and pilot testing of a nutritional curriculum for mid-adolescents. *Journal of Cancer Education, 8*(2), 145-150.

Weber, M. (1947). *The theory of social and economic organization* (A. M. Henderson & T. Parsons, Trans.). New York: Oxford University Press.

Weiner, R. L., Pritchard, C., & Frauenhoffer, S. M. (1993). Evaluation of a drug-free school and community program. *Evaluation Review, 17*(5), 488-503.

Weiss, C. (1995). Nothing as practical as good theory. In T. P. Connell, A. C. Kubisch, L. B. Schorr, & C. Weiss (Eds.), *New Approaches to Evaluating Community Initiatives, Vol. 1: Concepts, Methods, and Contexts* (pp. 65-92). Washington, DC: Aspen Institute.

Weiss, C. H. (1997). How can theory-based evaluation make greater headway? *Evaluation Review, 21*(4), 501-524.

Weiss, C. (1998). *Evaluation* (2nd ed.). Englewood Cliffs, NJ: Prentice Hall.

Weiss, C. H., & Bucuvalas, M. J. (1980). *Social science research and decision-making.* New York: Columbia University Press.

Weisz, J. R., Weiss, B., & Donenberg, G. R. (1992). The lab versus the clinic: Effects of child and adolescent psychotherapy. *American Psychologist, 47,* 1578-1585.

Wholey, J. S. (1979). *Evaluation: Promise and performance.* Washington, DC: Urban Institute.

Wholey, J. S. (1987). Evaluability assessment: Developing program theory. In L. Bickman (Ed.), *Using program theory in evaluation* (New Directions for Program Evaluation, No. 33, pp. 77-92). San Francisco: Jossey-Bass.

Wholey, J. S. (1994). Assessing the feasibility and likely usefulness of evaluation. In J. S. Wholey, H. P. Hatry, & K. E. Newcomer (Eds.), *Handbook of practical program evaluation* (pp. 15-39). San Francisco: Jossey-Bass.

Witkin, B. R., & Altschuld, J. W. (1995). *Planning and conducting needs assessments: A practical guide.* Thousand Oaks, CA: Sage.

Witte, K., Meyer, G., & Martell, D. (2001). *Effective health risk messages.* Thousand Oaks, CA: Sage.

Yin, R. K. (2003). *Case study research design and methods* (Applied Social Research Methods Series Vol. 5, 3rd ed.). Thousand Oaks, CA: Sage.

INDEX

ABOUT THE AUTHOR

Dr. Huey-Tsyh Chen has been a professor at the School of Public Health in the University of Alabama at Birmingham since 2002. Dr. Chen worked at the University of Akron until 1997, when he joined the Centers for Disease Control and Prevention (CDC) as chief of an evaluation branch. He took a leadership role in designing and implementing a national evaluation system for evaluating CDC-funded HIV prevention programs based in health departments and community-based organizations. Dr. Chen has contributed to the development of evaluation theory and methodology, especially in the areas of program theory, theory-driven evaluations, and evaluation taxonomy. His book *Theory-Driven Evaluations* has been recognized as one of the landmarks in program evaluation. In 1998, he received the Senior Biomedical Research Service Award from the CDC. He is also the 1993 recipient of the Paul F. Lazarsfeld Award for contributions to Evaluation Theory from the American Evaluation Association.